THE VEDIC FATHERS OF GOELOGY

ॐ

THE VEDIC FATHERS OF GOELOGY

By

NARAYANA BHAVANRAO PAVGEE

AUTHOR OF "THE BHARTIYA SAMARAJYA OR HINDU EMPIRE" IN 22 VOLUMES IN MARATHI

COSMO PUBLICATIONS

2001 INDIA

THE VEDIC FATHERS OF GEOLOGY
First COSMO Print 2001

ISBN 81-7755-111-6

Published by
MRS. RANI KAPOOR
for COSMO PUBLICATIONS Div. of
GENESIS PUBLISHING PVT. LTD.
24-B, Ansari Road, Darya Ganj,
New Delhi-110 002, INDIA

Printed at
Mehra Offset Press

DEDICATED

With the Deepest Respects

AND

The Utmost Gratitude

TO

Shree Sad-Guru

CONTENTS.

	Page.
INTRODUCTION.—	1—X.
CHAPTER I.—The Vedic Discoveries in Geology.	1- 24
CHAPTER II.—The Great Antiquity of the Vedas from the Geological point of view.	25- 85
CHAPTER III.—Modern Geological Researches of the West.	86- 97
CHAPTER IV.—Geological Researches of the East and the West compared.	98–131
CHAPTER V.—An Episode of the Glacial Period, and the Subsequent Quaternary Era.	132–157
INDEX	159–182

Errata et Corrigenda.

Page.	Line	For	Read.
6	8	n roads	inroads.
6	25	Mans	Man.
7	20	Past-Vedic	Post-Vedic.
26	9	how	show.
29	20	वैंभ्यः	पूर्वेभ्यः
36	15	देशवतः परः	देशस्त्वतः परः
57	19	into the land	in the land.
63	12	al	all
65	25	and in	and they in
75	24	conquerring	conquering
81	20	next *p*. 82	next *pp*. 82-83
84	17	on	an
139	5	१२१-११	१२-१-११.
151	2	orioin	origin
151	12	Germa	German.
152	23	writes a	writes as.

श्रीसद्गुरवेनमः ।

INTRODUCTION.

It had struck me some time back, that our very ancient Forefathers that belonged to the Tertiary Period, as I would show later on, were not only the framers of the most wonderful language—the Sanskrit–but, were, as observed by the learned Professor Max-Muller,[1] 'the fathers of the most natural of Natural Religions, the makers of the most transparent of mythologies, the most genuine inventors of very subtle philosophies, the first givers of the most elaborate laws, the first workers in the construction of our fundamental concepts, the oldest ancestors that framed our first words, and the most ancient Poets of our thoughts.

Thus, they werc *tho acknowledged benefactors* of Mankind, and were therefore naturally supposed by the East and the West to be the

---

1 **"What can India teach us?" Edition 1883 *p*. 15.**

Repositories of varied knowledge and diverse arts and sciences, especially, as they had made great advance in almost all the important branches thereof, even in ancient times, and in the still more remote past.[2]

In the circumstances, I could not believe, that our fore-fathers of yore had practically left untouched the very important subject of Geology, or that they were totally in the dark, even as regards the elementary knowledge of the science.

And, it soon appeared to me, that there was no reason to rest merely on surmises,' as the

---

2 (*a*) *Vide* the Rig-Veda and the Vedic Literature, which, by the bye, belongs to a period extending from B. C. 2500 to the Tertiary Epoch. *vide* also Chapter II and V of this work, and *pp*. 26 @ 28, 72, 73, 74 @ 85, 135, 138, 155.

(*b*) Elphinstone says :— * * * " the early excellence of the Bráhmans in all these branches of learning," * * * (Elphinstone's History of India. *p*. 92. Second Edition).

(*c*) Sir William Jones says, " (India) which has ever been esteemed the nurse of sciences, the inventress of delightful and useful arts, the scene of glorious actions, fertile in the productions of human genius." * * * (Voyage to India).

(*d*) *Vide* Max-Muller's " What Can India Teach us?" Edition 1883. *pp*. 13, 14, 15, 21, 22, 109 @ 113, 250 @ 255.

whole mist had gradually vanished, and solid ground was, by and by, vividly perceived, which clearly showed that the Vedic Rishis had not only superficial knowledge of, but intimate acquaintance with Geology. For, while writing in Maràthi *The Bhâratîya Sámrâjya* or the History of the Hindu Empire, (in twentytwo volumes, of which, by the bye, eleven have already been placed before the Public, the remaining eleven having been in course of publication), as also another work in Maráthi and in English, *viz.*, *The A'ryávartic Home and the A'ryan Cradle in the Sapta-Sindhus*, or *From A'ryâvarta to the Arctic and from the Cradle to the Colony*, I had naturally to ransack the whole field of the Rig-Veda, and other Vedic and ancient Sanskrit Literature ; and it was then, that I had come across many passages, numerous comments, and sundry arguments. The study of these, first made me think, that our Vedic fore-fathers were, in all probability acquainted with the main features, and perhaps with even the minute details of Geology. I, therefore, gave a second thought to what had occurred to my mind as only a probability at first, and continued my researches with persistent energy. I then minutely examined the various original Sanskrit texts from this stand-point, carefully scrutinized them with the object of gaining some insight into the matter, and found *the probability* to be *a certainty*, as I thought that there were very cogent grounds and strong evidence to affirm, that the Ṛig-Vedic Ṛishis and subsequent sages had, in truth, very wide acquaintance with, and intimate knowledge of Geology, since, they, every now and then, referred to the Geological formations and incidents of the Azoic, Palæozoic, Mesozoic, Tertiary, and Quaternary Epochs, as also to the most prominent life-types thereof.

And here, it would not be out of place to remark, that the colossal Sanskṛit Literature *viz.*, Vedic, Purânic, and Classical, which forms an astounding monument of the everlasting and commanding genius of our ancestors, contains but stray passages in respect of the Earth's crust, its Geological formations, its fossils, its various life-types, in short, its palæontology; and I have ventured to make copious use thereof, as these only formed the sources of my information and the basis of evidence, on which the present superstructure has been built, no Geological works in Sanskṛit worthy of the name having been found, owing probably to the immense Sanskṛit Literature on a variety of subjects having been destroyed by foreigners and fanatics, who had neither sympathy nor taste for it[3]

The Reader, perhaps, will scout and pooh pooh the idea, by asking with an air of sneer and contempt, " What ! The Vedas and the Purânas contain *germs* of Geology !! Absurd !!!" But, with all deference, I would here request him to pause, to wait, and to exercise patience, before pronouncing his hasty judgment. For, as rightly observed by Professor Max Muller, " It is far easier and far more amusing for shallow critics to point out what is absurd and ridiculous in the religion and philosophy of the ancient world than for the earnest student to discover truth and wisdom under strange disguises." " The Sacred Books of the East are no longer a mere butt for the invectives of missionaries or the sarcasms of philosophers. They have at last

---

3 Says Mr. Vincent Smith in regard to this, as follows :—" The political decadence of Magadha never affected the reputation of the kingdom as the Centre and head-quarters of Buddhist learning, which continued to be cultivated sedulously at Nâland and other places up to the time of the Mahomedan conquest at the close of the twelfth century, when the monasteries with their well-stocked libraries were reduced to ashes." (The Early History of India. Second Edition. *pp.* 294-295). *Vide* also Tod's Rajasthan Vol. I. *p.* VIII, IX, XIV, 217. Edition 1880. Third Reprint.

been recognised as historical documents, aye, as the most ancient documents in the history of the human mind, and as palæontological records of an evolution that begins to elicit wider and deeper sympathies than the nebular formation of the planet on which we dwell for a season, or the organic development of that chrysalis which we call man." ("India. What Can it teach us ?" *pp*. 253,254. Edition 1883). I may, therefore, take this opportunity to remind the Reader, without fear of contradiction, that the Vedas contain many things not yet known to anybody, as they form a mine of inexhaustible literary wealth, that has only partially been opened, and has still remained *unexplored.*

Now, another objection also might be raised to the effect that, the Vedas may perhaps contain elements of Cosmology ; yet, Cosmology and Geology are two very different things altogether, and the one has nothing to do with the other. But, to this, a better reply cannot be given than in the words of Professor Judd, as he says, " The first who endeavoured to draw a clear line of demarcation between Cosmogony and Geology was Dr. James Hutton, who declared that Geology was in no way concerned 'with questions as to the origion of things.' But his doctrine on this head was vehemently opposed at first, and, though it has been continually gaining ground, it cannot even yet be said to be universally accepted." (*Vide The Student's Lyell*. Edited by John. W. Judd. *p*. 5. Edition 1896 ; and *pp*. 17, 23, 24, of this Work).

However, anticipating all the objections in respect of the question, I have endeavoured to clear the ground, as far as possible, even at the beginning of the first Chapter of this Work, and quoted acknowledged authorities in regard to every argument, in view of fortifying my conclusions. (*Vide pp*. 1 @ 4).

Here, I need hardly mention, that I am aware of the dangers that beset the student of Research, and shall therefore attentively bear them in mind, without losing sight thereof, in as much as, there is great chance of erroneous observation and consequent risk of incorrect inference from it. Morever, I am aware that one has also to keep himself perfectly free from all prepossessions and prejudices, when, in the very nature of things, just reasoning from careful study demands the sacrifice. And, having due regard to all these facts, I have endeavoured to continue my researches, with the essential care and assiduity which the subject ever demands.

These Researches, therefore, I beg to place before the *Public*, with not a little diffidence, in the following few chapters, and therefore, respectfully *crave their indulgence*, for the venture undertaken by me. For, I may here be allowed to observe, that in regard to facts stated in the sequel, I humbly believe, that I stand on solid ground, and shall, therefore, leave no stone unturned to establish and prove them, by producing evidence to strengthen my arguments. I, however, beg to remark, that the testimony which I promise to offer in proof of Vedic acquaintance with Geology, only represents but stray reminiscences and past impressions of our Vedic Fathers as regards the Science of Geology, in the absence of Geological works or Purànic, Geological treatises, which in the *ruthless foreign Vandalism*, appear to have been lost forever to the world.

Naturally enough, therefore, in the humble work placed before the Reader, there would creep errors of judgment, and there would perhaps, I am afraid, be much difference of opinion as well. But, at all events, nothing would be stated without authority, nor any the least attempt made to misrepresent facts, or pervert them to suit the desired end.

नामूलं लिख्यते किंचिन्नानपेक्षितमुच्यते ।

Now, for convenient treatment of the subject, I have divided this Work into five chapters as under :—

I. The Vedic Discoveries in Geology.

II. The Great Antiquity of the Vedas from the Geological point of view.

III. Modern Geological Researches of the West.

IV. Geological Researches of the East and the West compared.

V. An Episode of the Glacial Period, and the subsequent Quaternary Era.

Unpractised in the office and the art of systematic discussion of a scientific subject, I would certainly rejoice in my own incompetence, should it provoke the championship of abler hands, superior talents, and defter pens, on behalf of *Natural Justice*, due to our Vedic Fathers in the matter of Geology, against the conventional unfairness and unmerited slur that slanders them often.

I am aware of the sharp criticism and even deep resentment that I shall probably provoke. And yet, for the sake of filial duty to my *Primitive Ancestors* and *greatly Revered Vedic Fathers*, I shrink not from the encounter, nor desist from the undertaking, simply for fear of public stricture, derision, or contempt. For, in the atmosphere of freedom, free thought must needs be duly ventilated without any restraint, in the interest of Science and Research. I therefore, submit this my humble Work to the candid judgment of the Public.

Finally, I take this opportunity to state, that I have already endeavoured to place before the Public the original (now in the Press) of this Work in Marâthi—my *Mother Tongue*—and have now ventured to give, with a few alterations and additions, the same in English dress, in view of its being reviewed by the *Savants* of the East and the West.

Poona,
Dated 20th Feb. 1912.

N. B. Pâvgee.

THE VEDIC FATHERS OF GEOLOGY.

CHAPTER I.

The Vedic Discoveries in Geology.

Before actually entering upon the legitimate discussion of the subject in hand, I think it desirable to make a few observations in respect of the province and scope of Geology, in view of anticipating objections that may be raised, that the so-called geological discoveries, ascribed by me to the Vedic Fathers, do not in the least appertain to, or have any the remotest connection with, the purposes and methods, or objects and conclusions of Geology. I shall, therefore, endeavour to solve this point, and remove the doubts arising therefrom.

Now, Geology (from Gr. *ge* the earth, and *logos* a discourse or reasoning) embraces in its widest sense all that can be known or inferred of the constitution and history of the structure of our planet. Or, in the words of Sir Charles Lyell, an eminent geologist, "Of what materials is the Earth composed, and in what manner are these materials arranged ? These are the first inquiries with which geology is occupied".[1] Besides, researches into the

---

1 Lyell's Manual of Elementary Geology, 1855, *p.* 1.

history of animate creation also occupy geology, not to mention the discovery of facts substantiated by proofs, that "the external parts of the Earth were not all produced in the beginning of things in the state" in which we now actually see them, but that, on the contrary, they "acquired their actual configuration and condition gradually, under a variety of circumstances and at successive periods," during each of which, distinct kinds of plants and living beings flourished in waters as also on land, the remains of which are still found buried in the Earth's crust.

Professor John W. Judd, another geologist says that, "Geology is the science which investigates the successive changes that have taken place in the inorganic and organic kingdoms of nature. It inquires into the causes of those changes and the influence which they have asserted in modifying the surface and external structure of our planet." (The Student's Lyell. Edited by J. W. Judd, 1896. *p.* 1.).

Professor Charles Lapworth maintains, that the object of Geology is to examine the various materials of our planet, to describe the origin of rocks which build up the Earth's crust, to investigate the nature and modes of the formation of rocks, to mark the various changes they have undergone, to note the fossils and their order of appearance by means of proof afforded by the organic remains found in the stratified deposits, "to determine the sequence and meaning of the various geological formations," and to note the past mutations of sea and land.

Dr. James D. Dana, an American geologist, writes in respect of Geology, as follows :— "Geology .... (1) treats of the Earth's structure and its system of development, — .... its progress in rocks, lands, seas, and mountains, &c.; (2) its progress in all physical conditions, as heat, moisture, &c.; (3) its progress in life, or its vegetable and animal tribes." He moreover says, "Geology is sometimes defined as the science of the structure of the earth. But the ideas of structure and *origin* of structure are inseparably connected, and in all geological investigations they go together.... The science, therefore, is a historical science." (Manual of Geology. 1863 *pp.* 2-4).

To sum up. Geology embraces (1) a general survey of the Earth's surface-features, (2) a description of the rock-materials of the globe, (3) an account of the rocks in the order of their formation, including palæontology and gradual progress in life or evolution in vitality, and (4) a statement of the agencies that brought about geological changes.

In the light of these facts, therefore, there appear strong grounds to maintain, that our

Vedic Fathers had knowledge of Geology, as in the Vedic works, traces of the description of the Earth's structure and its origin, its vital gradation and progress, the commencement of life on Earth, the order of appearance and evolution of the several life-types, have been observed every now and then, and they seem to give an account,—crude and imperfect though it be, according to the knowledge that our Vedic ancestors then possessed,—of the agencies that produced the changes. Thus, the Vedic trend of thought, I need hardly say, reveals all these things in their true colour, and the same will be seen from the treatment of the subject in the following pages.

It seems from the priceless treasure of Vedic Literature, that our ancient Forefathers had incipient and yet scientific ideas in respect of the first Geological formations, and that they had certainly made *crude but correct beginnings*, in the investigations of the science.

Probably, the first attempts having been crowned with success, a fresh stimulus was afforded to wider researches in the field, which, in time, having secured a greater share of the legitimate fruits thereof, marvellous results were gradually achieved, even in those so-called prehistoric times.

I admit, that in the important branch of the Science like Geology, no work, nor any treatise, nor for the matter of that any systematic collection on the subject, is now forth-coming to indicate any the remotest acquaintance with the science. But, with all these apparent short-comings, there are ample materials, abundant stray notes, and fragmentary details, scattered here and there, and probably throughout the vast field of Vedic Literature, from which inference can be reasonably drawn, that our Vedic Rishis of the hoary past, not to say of the Tertiary[1] period, as I shall presently show, had deep knowledge of the subject.

---

1 For, Professor Judd says, "some observers have even maintained that there is evidence of his (Man's) existence in pre-Pleistocene times." p. 148. (The Student's Lyell. Edited by John W. Judd, 1896).

Besides, it has been ascertained beyond all manner of doubt, that works of art associated with extinct mammalia of the Tertiary period, (vide Sir Charles Lyell's Elements of Geology, 6th Edn. *p.* 273-274), in a cavern in Somersetshire. And the occurrence in a newly discovered cavê, called "Long Hole," of the remains of two species of Rhinoceros, in an undisturbed deposit, in the lower part of which were some well-shaped flint-knives, evidently of human workmanship, strongly corroborates and proves the existence of the Tertiary man.

But, apart from this fact, even Sir Charles Lyell, one of the greatest and acknowlědged authorities on the subject says that, "the two implements (of flint found in the valley of the Ouse, near Bedford) occurred at the depth of thirteen feet from the surface of

This being so, the query would naturally arise, that if the Vedic A'ryans had wide acquaintance with the elements of Geology, what is the reason of the paucity, if not the total absence, of any Indian Geological Literature worth the name. But, it is not necessary to go far in search of the same. For, during foreign -roads and desultory incursions, the unsympa-

---

the soil, and rested immediately on solid beds of oolitic limestone," (vide Lyell's Antiquity of Man, 4th Edn. *p.* 214), and that, "evidence obtained by Naturalists that some of the extinct mammalia of Minche Court really lived and died in this part of France, at the time of the embedding of the flint tools in fluviatile strata, is most satisfactory." (vide Lyell's Antiquity of Man, p. 172, Edn. 4th.).

Sir Charles Lyell has also remarked that, "two perfect (human) lower jaw with teeth (were found), all associated, in such a manner, with the bones of bears, large pachyderms, and ruminants, and so precisely resembling these in colour and state of preservation, as to leave no doubt in his (Lyell's Companion's) mind that man was contemporary with the extinct animals. (Lyell's Antiquity of Mans, 4th Ed. *p.* 72).

Now, in connection with the aforesaid human implements of flint, found resting on solid beds of oolitic limestone, I might here quote with advantage Sir Charles Lyell, as he says, in his Elements of Geology, p. 92, 6th Edn. that "the strata above the chalk have been called Tertiary," and all this proves beyond doubt the existence of the Tertiary Man, as the human implements were found rested immediately on solid beds of oolitic lime-stone.

thetic invaders had consigned to the flames, and reduced to ashes, immense libraries in various parts of India, and the scenes of devastations, massacres, and plunders, which lasted through ages, have only served to revive the memory of the past. In short, after almost every city and capital was stormed and repeatedly sacked by barbarous foes, ruthless enemies, and exasperated plunderers, nearly all that was sacred in religion, every thing that was devoted to science, and whatever was but wonderfully remarkable in art, was destroyed without compunction and without remorse. It is, therefore, too much to expect that the literature of the country should not have suffered irretrievable loss,[1] by the wholesale destruction of valuable productions of men of genius, of genuine scientific works, and of useful arts and industry. Hence, evidently, the dearth and absence of Vedic or Past-Vedic Geological Literature.

With these requisite preliminary observations, I shall proceed to investigate the subject

1 Vide the Author's *Hindu Empire* or *Bhâratîya Sâmrâjya*. Vol. iii. Introduction, and Chapter XV.

In connection with this sort of Vandalism, it is also very painful to note, that the great and wonderful Library of Nálanda which contained a considerable stock of very interesting productions of Science and Art, of ancient literature and wisdom, was reduced to ashes, as noticed before. (Vide Introduction, Foot-note 3).

in hand. Before, however, dealing with the Vedic Literature, which abounds in allusions to a variety of distinct geological formations of different periods, I shall briefly refer to the Code of Manu, the Puránas, and the Mahâ-Bhárata, as they contain fragmentary details of, and garbed allusions to, geological rocks.

Going back first to the times of Manu, the great Hindu Lawgiver, it will be perceived how he has described the primary formations of our Planet. For, he says, that all this was, at first, in a chaotic state. The Heavens then created waters, and the Earth became enveloped therein. Subsequently, life was produced in it, and it gave rise to a variety of seeds or progressive vitality.

असीदिदं तमोभूतमप्रज्ञातमलक्षणम् ।
अप्रतर्क्यमविज्ञेयं प्रसुप्तमिव सर्वतः ॥ ५ ॥
अप एव ससर्जादौ तासु बीजमवासृजत् ॥ ८ ॥
(मनुस्मृतिः १-५।८).

As to the age[1] of Manu, it appears that he flourished some three thousand years ago, or one

1 In regard to this, Elphinstone in his History of India says "This would make the author of the Code (Manu) live about 900 years before Christ. That the Code is very ancient is proved by the difference of religion and manners from those of the present times, no less than by the obsolete style." p. 438. 2nd Edn.

Another Historian Dr. Sir Wm. Hunter remarks, "Schlegel was confident that it (the date of Manu) could not be later than 1,000 B. C." (Indian Empire p. 114, 2nd Edn.).

thousand years before the Christian era, and yet his Code of Laws, even at the present day, governs mankind, counting over twenty-three crores, and coming under the appellation of Hindu, from the Himâlayas to the Cape Comorin, and the frontiers of Cabul to the Brahmaputrâ. Mr. John D. Mayne in his preface to the "Hindu Law and Usage," has described the Code of Manu as follows :—

"Hindu Law has the oldest pedigree of any known system of jurisprudence, and even now it shows no signs of decrepitude. At this day, it governs races of men, extending from Cashmere to Cape Comorin, who agree in nothing else except their submission to it." (Hindu Law and Usage. 1st Edition. *p.* IX).

Now, reviewing the text of Manu from the Geological stand-point, it will be seen that it affords us but meagre data, in regard to the Geological conceptions of our ancestors.

The Puránas also, it seems, had but a dim vision and not very clear idea of the early geological formations, as apparently they were not in touch with the geological investigations of the Vedic times. For, the Bhâgavata Purâna has stated to say, that the Earth in the form of an egg, was, at first, all enveloped in water, which, in course of ages, having produced vitality, was the source of innumerable life-types.

In the Vayu Purâna, however, a brighter vista is seen, and we find mentioned therein, that the Earth was at first but a mass of intense heat, which, having in time cooled down, was merged in and replaced by water; wherein after lapse of epochs, was produced vital creative energy.

आपोह्यग्रे समभवन्नष्टेऽग्नौ पृथिवीं तले । (vi. 1) &c.

In the Agni Purâna and Vishnu Purâna, moreover, references appear to have been made to the progressive development of vital gradation, viz. from the Palæozoic fish and the Mesozoic tortoise and the reptile, to the Tertiary mammalia, at last evolving Man—the Crowning piece of creation. But, in the Purânas, the description having been given in the grotesque ideas of incarnations and Mythological legends, the original scientific geological theory seems apparently to have been all lost in the dim mist of religion.

Now, in respect of the antiquity of some of these Purânas, Colebrooke–the great Oriental scholar—says, "Itihâsa and Purânas are anterior to Vyâsa," (Vide his Miscellaneous Essays, Vol. I. p. 11). Vyâsa was the celebrated author of the Mahâ Bhârata and the Purânas, and the half-brother to Bhîshma, the reputed Commander-in-chief of the forces of Kauravas, in the Great War, which seems to have taken place some five thousand years ago. For, says Dr. Bhândârkar, "It thus appears, that in the latter part of the sixth century, the war which forms the theme of the Mahâ Bhârat was considered to have taken place, about four thousand years before." (Antiquity of Mahâ Bhârata, Journal, Bombay Royal Asiatic Society. No. XXVIII. Vol. X 1871–72).

Professor Wilson, another Oriental scholar known to fame, also speaks of the genuineness and great antiquity of some of the Purânas, in the following terms :—"A very great portion of the contents of many (Purânas), some portion of the contents of all, is genuine and old." (Vishnu Purâna. Preface. *p*. VI).

Turning to the Maha Bhârata, we find in the latter frequent allusions made to the successive vital gradations, which distinctly show knowledge of, and acquaintance with, the various geological life-types of different epochs, ranging from the *Vindhyan* (विंध्यकाल) or *Pre-Cambrian* period of the Palæozoic vitality to the Miocene Era (शिवालिकयुग) of the Cainozoic or Tertiary times.

Thus, it seems, that during all these times, viz. after the Vedic period, our ancestors had apparently neglected to cultivate the science of Geology. They, therefore, could not be in touch with geological facts and details ; while want of research, coupled with incomplete investigations and ignorance of facts, only served to yield very meagre results in respect of the science.

The Upanishads, however, throw quite a flood of light on the geological discoveries of the times. While, the knowledge which the Rishis of the Rigveda period[1] show, in respect of the subject, is simply incredible, especially when we take into consideration the fact that the Rishis belonged[2] to the Tertiary period. I shall, therefore, endeavour to give to the reader some idea of the geological theories then in progress, or rather of a few stray-thoughts and notions,

---

1 It may not be out of place here to remark that, the Rig-Vedic period was evidently a part and parcel of the Tertiary Era, as we shall endeavour to prove in the next chapter ; and the latest discoveries and researches enable us to carry the antiquity of Man to this Epoch, by establishing the fact that he had existed even in the Tertiary Era.

2 In respect of this, Mr. Tilak says as follows :—

" The subject matter of these hymns (of Rig-Veda) is inter-glacial, though its origin is still lost in geological antiquity. (The Arctic Home in the Vedas. *p.* 457).

And again he says, " the ancestors of the Vedic Rishis lived near the North Pole in times before the last Glacial Epoch." (The Arctic Home in the Vedas. *p.* 464).

Sir Charles Lyell has also proved and shown, beyond all possibility of doubt, the existence of the Tertiary Man, in his well-known work—*The* Antiquity of *Man*. Fourth Edition. (Vide *ante p.* 5. Foot-Note).

which were at the time upper-most in the minds of the hoary Bards of the Vedic period, as they apparently seem to be but remnants of some scientific treatise, or of an exhaustive discussion on the subject, at the time.

It would, I think, be convenient for a wide survey of the subject, to begin with the Upanishads and give a few details, as the zealous Rishis of the period seem to have taken very keen interest in geological researches. From the substance of what has been stated in the Taittiríya Ṣamhitâ, it appears to have been supposed that at the beginning, all was liquid, which was probably considered to be enveloped in gaseous matter, and that this subsequently became transformed into the Earth. (T. S. VII. 1. 5. 1.). आपो वै इदमग्रे सलिलमासीत् । तस्मिन् प्रजापतिर्वायुर्भूत्वा अचरत् ....सा पृथिव्यभवत् । (तै. सं. ७-१-५-१).

The researches of the Taittiríya Brâhmaṇas go a step further, and declare that the Earth was at first but a burning globe of gaseous matter (तस्मात्तपनाद्धूमोऽजायत । ...अग्निरजायत ।) ; that it subsequently became liquid (स समुद्रोऽभवत् ।); and that (after it gradually cooled down, and became condensed and solid by the consolidated particles of granite), it was transformed into the Earth (सा पृथिव्यभवत् ।); and then vitality began to make its appearance in the form of lotus, (or leaves of moss and weeds, &c.). (T. Br. II. 2. 9. 1, 2, 3, 4 ; I. 1. 3. 5.).

I venture here to remind the reader, that religion having pervaded everything Hindu, or rather the whole Indo-Aryan life, the geological evolution and configuration of the Earth, the modification of its surface and its external structure, the formation of gneiss or gravels and the appearance of vitality in weeds or lotus, the Palæozoic vegetation and the fish, the Mesozoic tortoise, and the Tertiary mammalia, were all supposed to be the work of Prajâpati—the Creator—, as the following extracts will show :—

इदं वै अग्रे नैव किंच नासीत् । ... तदतप्यत । तस्मात्तपनाद्धूमोऽजायत । तद्भूयोऽतप्यत । तस्मात्तपनादग्निरजायत । ...ज्योतिरजायत । ..तदभ्रमिव समहन्यत । तद्वस्तिमभिनत । स समुद्रोऽभवत् । ...यदप्सु अवापद्यत सा पृथिव्यभवत् । (तै० ब्रा० २. २. ९. १, २, ३, ४.). सोऽपश्यत् (प्रजापतिः) पुष्करपर्णं तिष्ठत् । ...स वराहो रूपं कृत्वा उपन्यमज्जत् । स पृथिवीमधः आर्धत् । तस्या उपहृत्य उदमज्जत् । तां शर्कराभिरदृंहत् । तै० ब्रा० १. १. ३. ५). Here, वराह or the Boar represents the Mammalian life-types.

आपो वै इदमासन्सलिलमेव । सप्रजापतिरेकः पुष्करपर्णे समभवत् । ...योरसःसोऽपामन्तरतःकूर्मभूतं...तै० आ० (१-२३-१, ३).

The geological progress in the Taittiriya Upanishads is still more remarkable and surprising, as it shows the gradual evolution and vital gradation, with marvellous fertility of mind and astounding correctness of an expert. For, while describing the creation, as also the gradual evolution of Cosmos and the Earth, its cause and effect, the Rishis of the Upanishad-period say, that from the all-pervading Soul was produced the sky or the endless expanse of space, which in turn had produced the wind, the latter the fire, this the waters, these again the Earth, which (after it had cooled down) had given rise to vitality or vegetation, and the vital gradation having made its beginnings from simple herbs, plants, or weeds, had evolved more complex forms, ending in the crowning piece of creation, viz. the Man. (T. U. ii. 1).

तस्मादात्मन आकाशः संभूतः । आकाशाद्वायुः । वायोरग्निः । अग्नेरापः । अद्भ्यः पृथिवी । पृथिव्या ओषधयः । ओषधीभ्योऽन्नम् । अन्नात्पुरुषः । (तैत्तिरीयोपनिषद्. २-१).

Now, this description may, in brief, be said to be the geological evolution of Cosmos from the infinite space (आकाशः), which having given rise to atmosphere (आकाशाद्वायुः ।), the violent currents of wind produced fire (वायोरग्निः ।). This remained in a state of igneous fusion, until part of the heat having been diffused into the surrounding space, refrigeration proceeded in course of time, which having caused the aqueous vapour in the atmosphere to condense, occasioned the fall of rain (अभ्रेरापः ।), thus giving rise to what is called the first *Thermal Ocean.* The temperature, however, of this boiling sea was at first, in the very nature of things, very high, and therefore no aquatic beings could inhabit its waters. Consequently, it was devoid of fossils. But, the waters of the Thermal Ocean were highly crystalline, and hence the origin of the primary crystalline strata, gneiss, mica, gravels[1] (शर्करा), &c. Gradually, the granite crust[1] having been partially broken up,[1] land and mountains began to rise above the waters (अद्भ्यः पृथिवी ।), while the torrents of rain caused the sediments to spread over the bottom of the boiling sea. Now, the land and mountains that had appeared above waters having in time cooled down, along with the thermal fluid, life naturally became visible therein, first in herbs or seaweeds (पृथिव्या ओषधयः ।), and then in others, the vital gradation having progressed from simpler conceptions to more complex types and highly organised orders, ending in the last[2] and the marvellous creation of Man.[2] (ओषधीभ्योऽन्नम् । अन्नात्पुरुषः ।).

---

1 Vide ante page 14. (तां शर्कराभिरदृंहत् । तै० ब्रा० १-१-३-५.).

2 In respect of this, Professor Dana, the American geologist, says as follows : "It is not known that any new species of plants or animals have appeared on the Earth since the creation of Man." (Manual of Geology. p. 586, Edn. 1863).

I may here mention and venture to state, that the aforesaid description of the origin of things and of the Earth, in the Taittirîya Upanishad (ii. 1), is but a synoptical view of the gradual evolution of life-types and of geological formations, and corresponds, in the main and essential particulars, though not in so many words, to what the eminent geologists and acknowledged authorities of modern times say, as regards the existence and primary conditions of our Planet, the formation of its rocks, its mineral mutations, and its vital gradation and progress. And although the foregoing discussions have been held by some to apply to the evolution of Cosmos in Metaphysics, still, they, none the less, apply to geological rocks and formations, that have made their appearance on Earth, in consecutive sequence. Because, the same can be proved by ample evidence from the Rig-Veda and other sources, as we shall presently show.

As previously observed, (*p* 13), we find in the Taittirîya Brâhmana, allusion made to the creation. and its evolution, which in substance only means that the Earth was at first in a chaotic state and enveloped in gaseons matter (तस्मात्तपनाद्धूमोऽजायत), or was in a state of igneous fusion (तस्मापत्तनादग्निरजायत), which having] caused water to flow, after refrigeration and condensation of vapour, there was the *Thermal Ocean*

(स समुद्रोऽभवत्), and this subsequently gave rise to Earth (यदप्स्ववार्यत सा पृथिव्यभवत् । Taittiriya Br. II. 2-9-1, 3, 4).

After lapse of ages, the Earth gradually began to cool down, evidently by slow and continuous process, the vast effects of which could only have been realized after period of almost incalculable duration. And weeds and lotuses (पृथिव्या ओषधयः T. U. II. 1 ; सो ऽपश्यत् पुष्करपर्णं तिष्ठत् T. Br. I. 1-3-5), trilobites and fishes, reptiles and quadrupeds, (अकरोत्स तनूमन्यां कल्पादिषु यथा पुरा । मत्स्यकूर्मादिकं तद्वद्वाराहं वपुरास्थितः । Vishnu Purâna I. 4-7), began to swarm and breed, on earth. It was supposed in those stirring intellectual Vedic times, that life was dormant even in the fluid, when the Earth was in a liquid state. For, it appears from the Taittirîya Áranyaka (I. 23-3/4), that the fluid became a tortoise moving amid the waters (यो रसः सो ऽपामन्तरतः कूर्मं भूतं सर्पन्तमब्रवीत् ।), and that when Prajápati or the Creator said to him, " *Thou hast sprung from my skin and flesh,* " the tortoise had then replied to say, " *No. I was here even before.* " (नेत्यब्रवीत् । पूर्वमेवाहमिहासमिति ।). This, therefore, evidently indicates that life was supposed to be dormant even in the fluid of the deep.

Proceeding further, and going back to the very root of things and the foundation of all, we come to the Rig-Veda, which is admitted the oldest[1] document in the world, and which claims civilization the most ancient. Here, then, reading between the lines the Taittirîya Brâhmana, (ante *p* 14), we find it indicated by Vedic Rishis of the hoary past, that the Earth, having been at first in a cloudy, vapoury, or gaseous state (of igneous fusion), was reduced to the molten condition of water, and that it (the Earth) having been thus in a liquid condition, and found trembling or rolling like water (पृथिवीं व्यथमानां), and the mountains also moving to

---

1 In respect of this, Mr. B. G. Tilak says, " The Vedas themselves.........admittedly form the oldest records of the Aryan race." (The Arctic Home in the Vedas. *p* 5).

Professor Weber writes, "We are fully justified in regarding the Literature of India as the most ancient literature, of which written records on an extensive scale have been handed down to us." (History of Indian Literature. 1882. *p*. 5).

In the same way, Professor Max-Muller contends and declares the Rig-Veda and the other Vedas to be the "most ancient literary documents." *p*. 116. (What can India teach us ? Ed. 1883).

He also observes that, " The Vedic Poets are primitive, the Vedic language is primitive, the Vedic religion is primitive, and taken as a whole, more primitive than anything else, that we are ever likely to recover in the whole history of our race." *p*. 124. (What can India teach us ? Edition 1883).

and fro like waves, (having evidently been in a molten state पर्वतान् प्रकुपितान्), Indra made them firm (स प्राचीनान्पर्वतानदृंहत् । R. V. II. 17. 5).

Thus, we find that he made the Earth solid (पृथिवीमदृंहत्), and the mountains firm (पर्वतानरम्णात्). The verses from the Rig-Veda being very important, I give them here below, for ready reference :—

यः (इंद्रः) पृथिवीं व्यथमानामदृंहत्
यः पर्वतान्प्रकुपितानरम्णात् ॥ (R. V. II. 12. 2.).
स प्राचीनान्पर्वतानदृंहत्.. ....
ओजसा ऽधराचीनमकरोदपामपः ॥ (R. V. II. 17.5).

This may be freely rendered into English as follows :—" Indra made the Earth solid that was trembling (having been at first in a molten state or condition of watery vapour). He also made the mountains firm that were rolling." (R. V. II. 12. 2). " He made the waters flow from the vapoury clouds, thus reducing by his might the gaseous state of Earth to the molten condition, and rendered the mountains firm." (R. V. II. 17. 5). In fact, the verses when read along with T. Br. II. 2. 9. 1; I. 1. 3. 5, *p* 14, mean, although not distinctly expressed in so many words, that the Earth was at first in a gaseous state of igneous fusion ; that subsequently, part of the heat thereof having been diffused in the surrounding space, refrigeration had gradually proceeded ; that this naturally caused the aqueous vapour in the atmosphere to condense, occasioning thereby the fall of rain, and thus creating the *Thermal Ocean* ; and that then, the granite crust having been formed and afterwards partially broken, the land and mountains began to appear in time. (Vide ante *pp* 15-16).

In another place also, mention is made again of the Earth having been made solid by Indra (येन पृथिवीं च दृढा । Rig-Veda. X. 121. 5.).

But, more than this, and as if to crown the whole, we find a most important statement made in the Rig-Veda, from the distinct standpoint of Geology, that vegetable life (ओषधीः) had commenced (पूर्वा जाता), some three epochs (त्रियुगं), before the Mammalian life-types of the Tertiary Era, or rather before the advent of Man and the superhuman Gods (देवेभ्यस्त्रियुगं पुरा) ; and that in the Primary[1] or Palæozoic Era (पूर्व्ये युगे), vitality had first come into being, after the Azoic Epoch (असतः सदजायत ।). That is to say, from the chaotic deep or the *Thermal Ocean*, where nothing was favourable to the sustenance of life, there was the manifestation of vitality, after the boiling waters had cooled down. In other words, it was then, that life had come into play and existence become

1 So called by Sir Charles Lyell–the great Geologist–in his Elements of Geology. Sixth Ed. p. 92.

manifest from *unmanifested nothing* (असत: सद्जायत ।). I give below the original *in extenso*, with the translation as rendered into English by Griffith.

या ओषधी: पूर्वा जाता
देवेभ्यस्त्रियुगं पुरा । (Rig-Veda. X. 97. 1).
देवानां पूर्व्यें युगे
ऽसत: सद्जायत ॥ (Rig-Veda. X. 72. 2).

"Herbs that sprang up in time of old, three ages earlier than the Gods." (Rig-Veda. X. 97. 1).

"Existence, in an earlier age of Gods, from non-existence sprang." (Rig-Veda. X.72. 2).

This certainly very favourably compares with the keen investigations of the modern and reputed geologists of the West, as they affirm that vitality had first commenced with the herbs—(the Algæ, or sea-weeds)—in the Primary[1] or Palæozoic Epoch, followed next by the Secondary or Mesozoic Period or the Age of Reptiles, the latter having in turn been succeeded by the Tertiary Era, that is, Cainozoic or the Age of Mammals and of Man, of which, however, I shall give details in Chapter IV.

---

1 Judd says that, "the use of the term Primary is now almost entirely abandoned." (The Student's Lyell. p. 127. Ed. 1896). Lyell, however, has used it. Vide ante p. 21, Foot-note.

Now, it might perhaps be argued, that the various theories propounded by the Vedic and Purânic Indians, either in the Purânas and the Mahâ Bhárata, the Upanishads and the Rig-Veda, were advanced not with reference to Geology, but for the sole purpose of Cosmology. And to this, a curt reply might with advantage be given, that whatever the purpose, if the results achieved are in the main correct, even from the scientific point of view, the researches made by our hoary ancestors, in the dim and distant past, must needs be appreciated. But, apart from this, we have distinct evidence to show, that the Vedic Rishis had wide acquaintance with Geology, and that they had made these statements especially with reference to geological formations and geological facts. This, however, requires a separate treatment. We shall, therefore, discuss it in the fourth chapter, while investigating the progress of the West in the science, and comparing it with our Vedic Researches in the past, in this very interesting subject of Geology.

In the circumstances, and having had due regard to the great antiquity of the Vedas from the geological point of view, of which I shall endeavour to give some idea to the reader, in the ensuing chapter, the above exposition made by the Vedic Rishis, of the first geological for-rations, appears simply marvellous, owing not only to its great antiquity, but also to its correct demonstration, even according to, and from the stand-point of, the discoveries of modern researches and the accepted opinions of eminent Geologists of the West, who had taken up the subject only over two centuries before. This, therefore, I shall notice in a separate chapter, as the Western Geologists have rendered valuable assistance to the cause of the Science of Geology, and have developed it with untiring zeal, great energy, and commendable perseverance, after giving to the Reader a brief idea about the antiquity of the Vedas from the stand-point of geology, in the next chapter.

CHAPTER II.

The Great Antiquity of the Vedas from the Geological point of view.

In the present chapter, I intend conveying to the Reader some idea about the antiquity of the Vedas, and especially of the hoary past of the Rig-Veda, to enable him to see how the geological knowledge possessed by the Vedic Rishis, very favourably compares with that of the modern Geologists of the West, notwithstanding the vast distance of time and the almost insuperable difficulties that the Rishis had then to encounter in their geological researches.

From the foregoing pages, the Reader will have at once perceived, that ancient Sanskrit Literature abounds in allusions to the various Geological formations, while the Vedas, every now and then, refer to different rocks and vital gradations, in Geological succession and gradual sequence of Evolution.

Of the several Vedas, the Rig-Veda is pre-eminently the most ancient, and admittedly the oldest document in the world, of which, therefore, we shall speak presently, as it furnishes unquestionable evidence in respect of the geological knowledge of the Vedic Rishis in prehistoric times. But, before we do that, it would, I think, be proper to give here the approximate date of the Upanishads and the Brâhmanas, which, as we have already observed, are replete with allusions to the various rock-formations of considerable importance.

The researches of the Oriental Scholars show, that the period of the Upanishads and the Brâhmanas was over 2,500 years before A. D., while the period of the Rig-Veda extended even beyond the Quaternary Era. That is to say, many hymns of the Rig-Veda, if not all, were sung by the Vedic Rishis during the Tertiary Epoch, preceding the Great Ice Age, as would appear clear from the evidence offered in the sequel. Besides, the great antiquity of the Vedic period has also been recently acknowledged by acute researchers like Messrs. B. G. TILAK, S. B. Dixit, V. B. Ketkar, &c. of the East, and critics like Dr. Jocobi of Bonn, Dr. Buhler, Professor Bloomfield, Professor Max-Muller, and others of the West.

Now, as regards the Brâhmanas and their ascertained period, Mr. Tilak says, that, " the astronomical statements found in the Vedic Literature supplied us with far more reliable data for correctly ascertaining the ages of the different periods of Vedic Literature, " and that these statements " unmistakably pointed out that the vernal equinox was in the constellation of Mriga or Orion (about 4,500 B. C.), during the period of the Vedic hymns, and that it had receded to the constellation of the Krittikas, or the Pleiades (about 2,500 B. C.) in the days of the Brâhmanas." He further states in his very interesting work *The Arctic Home in the Vedas* that, "the subsequent discovery," by his friend the late Mr. S. B. Dixit, "of a passage[1] in the Shatapatha Brâhmana, plainly stating that the Krittikas never swerved in those days, from the *due East*, that is, the Vernal Equinox, has served to dispel all lingering doubts regarding the age of the Brâhmanas; while another Indian astronomer, Mr. V. B. Ketkar, in a recent number of the Jonrnal of the Bombay Branch of the Royal Asiatic Society, has mathematically worked out the statement in the Taittiriya Brâhmana, (III. 1, 1, 5)." (Vide Mr. Tilak's *Arctic Home in the Vedas*. Preface. *pp* I. II.).

I here give below, the text from the Taittirîya Brâhma*na*, and this purports to mean that, "Brihaspati or the planet Jupitor was first discovered when confronting or nearly occulting the Star Tishya"

---

१ एताः (कृत्तिका) ह वै प्राच्यै दिशो न च्यवन्ते । सर्वाणि ह्वा अन्यानि नक्षत्राणि प्राच्यै दिशश्चयवन्ते । (श० ब्रा० २-१-२-३).

बृहस्पतिः प्रथमं जायमानः। तिष्यन्नक्षत्रमभिसंबभूव।
(तै० ब्रा० ३-१-१-५).

Evidently, the observation of the phenomena is a relevant fact, and as it has been proved by the text itself, Mr. Ketkar has endeavoured to establish, that this sort of observation of the heavenly bodies was possible only at about 4650 B. C.

But, apart from this, Mr. Tilak, resting on other grounds, which he has very ably discussed in the Orion, has stated to say, that " the oldest period in the Aryan Civilization may therefore be called the Aditi or the pre-Orion period, and we may roughly assign 6000 – 4000 B. C., as its limits ". (Vide *The Orion*, or the Researches into the Antiquity of the Vedas. *p* 206, 1893).

However, having had due regard to the matter-of-fact evidence found in the Rig-Veda, which goes to prove the great antiquity of the Rig-Veda,—antiquity going so far back as the hoary period of the Tertiary Epoch, of which therefore I shall speak later on,—one is likely to ask whether the aforesaid period of 6000 – 4000 years B. C., is not too modest for the oldest Vedic age. And, here, therefore, I am constrained to think that, the aforesaid limit is by no means to be considered as the *Ultima*

Thule of the Vedic antiquity, in as much as the Aryan Civilization and the primitive Vedic life appear to have preceded the Vedic Antiquity by thousands of years. For, the oldest Vedic Rishis themselves speak in the Rig-Veda of their most ancient ancestors and of the by-gone ages, as also of their then modern generation, as will be seen from the following :—

अग्निः पूर्वेभिर्ऋषिभिरीड्यो नूतनैरुत ।
(ऋ० वे० १-१-२).

" Agni is the God that deserves to be praised by Rishis or Sages, ancient as well as modern."

येषां पूर्वेषामश्रृणो ऋषीणाम् ।
(ऋ० वे ७-२९-४

" Oh Maghavan (Indra), you had listened to the praises of our older ancestors."

इदं नम ऋषिभ्यः पूर्वजेभ्यः पूर्वेभ्यः पथिकृद्भ्यः ।
(ऋ० वे० १०-१४-१५).

" Bow to our older ancestors that led us to good path, and to the Rishis born before the creation"—(सृष्ट्यादावुत्पन्नेभ्यः as explained by the great commentator Sáya*n*a).

This being the case, it is no wonder that the trend of thought of western Scholars should also be in the same direction, and it is in the very nature of things that they should have been inclined to admit that the Vedic antiquity must have been preceded by the hoary past. For, the facts are self-evident, and they themselves prove the conclusion. It was for this reason, that Professor Bloomfield, while reviewing Mr. Tilak's *Orion* or *the Researches into the Antiquity of the Vedas*, had, in his address on the occasion of the Eighteenth Anniversary of John Hopkin's University, very rightly observed that, " The language and literature of the Vedas is, by no means, so primitive as to place with it the real beginnings of Aryan life. " Inasmuch as, he says that, " these, in all probability, and in all due moderation, reach back several thousands of years more," and naturally, therefore, he argues that, it was " needless to point out, that this curtain, which seems to shut off our vision at 4500 B. C., may prove in the end a veil of thin gauze."

And this gives rise to a series of questions, which, therefore, cannot be left unnoticed :—

(1) How old are the Vedas, and to what Geological Epoch does their antiquity extend ?

(2) Were our (Indo-Aryan) ancestors of the Rig-Veda times older than the Quaternary Period ?

(3) Did they belong to the Tertiary Era ?

(4) Had they seen the Great Ice-Age ?

(5) If so, is there any evidence in the Rig-Veda or other Vedic and Avestic works, to support the statement and fortify the conclusions ?

(6) Is there any scientific evidence to prove the existence of the Tertiary Man ?

It will be perceived that an answer to the last question would serve as a key to the solution of the second, the third, and the fourth. I shall, therefore, try to do justice to it first, in view of making our task simpler and easier, in respect of the rest of the aforesaid queries.

Now, Messrs. Medlicott and Blanford of the Indian Geological Survey by the Government of India make the following observations in respect of the Miocene deposits and Shiwâlik Mammalia of India :—" The valley gravels of the Indian Peninsula, and especially some fossiliferous beds in the Narbadà valley, contain a few Shiwâlik Mammalia, associating with species more nearly allied to those now living. Remains of human implements have also been detected in these gravels, which are probably of Post—Tertiary or Pleistocene age." (Vide Manual of the Geology of India. *p LV*).

But, the fact that man has been in existence from the Tertiary Period seems now to be an acknowledged fact, as the same has been proved by scientific evidence. For, Professor Ernst Hæckel of Germany says, " The first appearance of man, or to speak more correctly the development of man from the most nearly allied apeform dates probably either from the Miocene or the Pliocene period, from the middle or the latest section of the Tertiary Epoch." (Evolution of Man. Vol. II. *p* 16. 1879).

Dr. Noetling also had, in 1894, discovered the works of Pliocene Man in Upper Burmah. (Keane's Ethnology. *p* 454. Edition 1899).

Moreover, Professor Keane has stated in his other work " *Man Past and Present* " that, " Flints " (worn and polished by human action) have been " found *in Situ* associated with the remains of such extinct fauna as Rhinoceros * * * assigned to the Lower Pliocene. " (*p* 5. Ed. 1899).

Dr. Frederick Wright also maintains, that relics of Man are found in the Tertiary Epoch, and declares that, " the expectation of finding evidence of Pre-Glacial Man in Ohio was justified soon after this," that is, in 1855. (Vide his work " Man and the Glacial Period. " *p* 249. Ed. 1892).

And above all, even Sir Charles Lyell, the great authority in Geology, has admitted the existence of the Tertiary Man, from the facts available and the evidence placed before him, which he had personally scrutinized and sifted, and very ably discussed in his interesting work entitled " The Geological evidences of the Antiquity of Man ", with all the requisite care which the subject demanded. (Vide ante *p* 5. Note).

Besides, relics of Miocene Man were also found in Further India, and, therefore, in regard to this, Edward Clodd, President of the Folk Lore Society stated that, " Quite lately, there have been discovered in an Upper Miocene deposit in Further India, some clipped flint flakes, of undoubted human workmanship. " (The Story of Primitive Man *p* 23. Ed. 1895).

Thus, the Tertiary Man having been proved, the solution of the second and the fourth question has naturally become simpler. The answer thereto is evidently in the affirmative, since both the Ice Age and the Quaternary Era were preceded by the Tertiary Epoch : while, our Rig-Vedic Rishis had seen the last and the Great Ice Age, and were older than the Quaternary period, as we shall presently show.

Now, we shall for a while turn our attention to the remaining questions, and see if there is any Vedic evidence in support of the facts alleged, and referred to above. In my work entitled, "*The Âryâvartic Home and the Aryan cradle in the Sapta Sindhus*", or "*From Âryâvarta to the Arctic and from the Cradle to the Colony*", and in my larger work in Marathi with still greater details, (both now in the Press), I have endeavoured to prove, by all sorts of evidences, Vedic and non-Vedic, Scriptural and profane, scientific and demonstrative, historical and traditional, that we are autochthonous in India; that we were born in Âryâvarta on the banks or in the region of the reputed and the most sacred river the Sarasvati, which was deemed by our very ancient Vedic ancestors of the Tertiary Period to be the scene where life had first commenced; that our Colony of young adventurers, having emigrated from and left Âryâvarta, had colonised distant lands of Asia, Africa, Europe, and America, and settled in the Arctic and Circum-Polar regions, during the Tertiary Epoch, at a time when the climate of the Arctic regions having been genial, these were fit for human habitation; that at the sight of the new phenomena of everlasting Dawns, as also of the unusual long days and nights of the Arctic Regious,—to which our colonists from India (Bhârata-Varsha) were not accustomed while living in their Mother-Country-Âryâvarta,–their astonishment and fear knew no bounds ; and that at the advent of the great Ice-Age, the once genial climate of the Arctic Regions having been replaced by extreme, not to say unbearable cold, and the higher latitudes having been covered with Ice-caps of enormous thickness, such of our colonists as had made settlements there, were compelled to retrace their steps back to their Mother-land Âryâvarta, by the direction of the Snow-clad Himâlaya, which was ever in their minds, and which they always remembered and cherished with fondness, as the *Northern Boundary* (उत्तरं गिरिं) of their Beloved Bhârata-varsha. I shall, therefore, venture to recapitulate some evidence here, in brief, to save reference, for our present purpose.

Manu, our very ancient, famous, and well-informed Law-giver (vide ante *p.* 8), has in his मनुस्मृति or the Code of Laws, declared Brahmâvarta (ब्रह्मावर्त) to be *the God-created region*, situated in Aryâvarta and between the two *divine* rivers, the Sarasvatî and the Drishadvatí (II. 17). Evidently, this was supposed to be *the scene of creation* (vide Muir's *Original Sanskrit Texts*, vol II. *p* 400. Second Revised Edition), and the pronouncement appears to have been made by Manu, not in the least without strong grounds. For, it surely rests upon solid facts, traditional evidence, and even Vedic authority, which, therefore, we shall proceed to examine presently.

In the first place, the traditional impressions seem to have been current and even engraved on the hearts of the Indian Aryans of the Manu-Period, that Âryâvarta has been *their cradle* ; that *this region was the source of pure usage* (सदाचारः), handed down from generation to generation (पारंपर्यक्रमागतः । म० स्मृ० २-१८) ; that as such, it was even thought worthy of being copied and learnt by other nations on Earth, from the Brâhmans (एतद्देशप्रसूतस्य सकाशादग्रजन्मनः । स्वं स्वं चरित्रं शिक्षेरन्पृथिव्यां सर्वमानवाः ॥ म० स्मृ० २-२०) ; and that the country beyond the limits of *the sacrificial* Region, viz. Âryâvarta, belonged to *foreigners* (म्लेच्छदेश यतः परः । म० स्मृ० २-२३).

Moreover, if at all, we ourselves were foreigners in Âryâvarta, and had immigrated into the Land of the Seven Rivers, Manu certainly would never have said that, "the country beyond the confines of Âryâvarta was of the Mlechhas." Because, evidently enough, there could have been no propriety in using the expression.

Now, as to the further traditional testimony, I may here briefly observe, that it is not only the Hindus or rather the Indo-Aryans that think themselves to be autochthonous in India in consequence of the hoary traditions received from father to son, but it is also the foreigners that consider

India as the *cradle* of the Hindus. For, while discussing this question, Elphinstone in his History of India says : " It is opposed to their foreign origin, that neither in the Code (of Manu), nor, I believe, in the Vedas, nor in any book that is certainly older than the Code, is there any allusion to a prior residence, or to a knowledge of more than the name of any country out of India. Even mythology goes no further than the Himâlaya chain, in which is fixed the habitation of the gods." (History of India. Vol. I. *p*. 97. Second Edition).

Besides, in regard to the deep-rooted traditions of us—Indians—in respect of A'ryâvarta having been our *cradle*, I cannot do better than quote the words of the great commentator. For, Kulluka is the name of the exegetist who has written his commentory on the Code of Manu, and he declares in no equivocal language to the effect, that " Aryâvarta is the *region* where Âryans *were born*, and *are born*, and *re-born*, *over and over again*." The original in Sanskrit, under verse 22, Book Second, of Manu's Code, runs as follows :—

आर्या अत्रावर्तन्ते पुनः पुनरुद्भवन्तीत्यार्यावर्तः ।

But, more than this, even before and at the time of Mahâ-Bhârata, the idea having taken a deep root, had a firm hold on the

Indian mind, that the region lying between the two divine rivers, *viz.* the Sarasvati and the Drishadvati, was *the scene of creation*, as it was *the tract fashioned by God.* In fact, in Mahâ-Bhârata, it has been called *Brahmâvarta* (देवनिर्मितं देशं ब्रह्मावर्तं प्रचक्षते । म० भा० १४-११०-४४). Now, the expression god-created region, or *the tract fashioned* by God, appears to have had a very great significance, and it emphatically presents to us the *deep-rooted* traditions of *Our A'ryan Cradle* having been in Brahmâvarta, and as such in Âryâvarta alone, and nowhere else. There is another verse in the Mahâ-Bhârata, which is also very interesting. For, this says that, "the region has its own *hoary traditions and customary usages*" (यस्मिन्देशे य आचारः पारंपर्यक्रमागतः ।), handed down from age to age and transmitted from generation to generation, which, therefore has been supposed to be *the only pure usage* (स सदाचार उच्यते । Mahâ-Bhârata. XIV. 110. 45). But, above all, there is yet a third verse, and this is certainly pregnant with meaning, inasmuch as it declares that *Brahmarshi-Desha*, or for the matter of that Âryâvarta, is *a model region of discipline and order*, and that "from a Brâhman born in that region, all men on earth (पृथिव्यां सर्वमानवाः ।) should receive *their lessons* in their respective *duties* and responsibilities."

(सकाशादग्रजन्मनः । स्वं स्वं चरित्रं गृह्णीयुः । M. Bh. XIV. 110. 47). For easy reference, I herein below quote the verses from Mahá-Bhârata, and these as also others seem, with a few variations, to have been repeated by Manu in (II. 17, 18, 20), for confirming the hoary traditions in respect of our Cradle in Âryâvarta).

सरस्वतीदृषद्वत्योर्देवनद्योर्यदन्तरम् ।
तद्देवनिर्मितं देशं ब्रह्मावर्तं प्रचक्षते ॥ ४४ ॥
यस्मिन्देशे य आचारः पारंपर्यक्रमागतः ।
वर्णानां सान्तरालानां स सदाचार उच्यते ॥ ४५ ॥
एतद्देशप्रसूतस्य सकाशादग्रजन्मनः ।
स्वं चरित्रं च गृह्णीयुः पृथिव्यां सर्वमानवाः ॥ ४७ ॥

(श्रीमन्महाभारते अ० ११०).

(The South Indian Texts Edition).

Turning to the Vedic authority for a while, it seems that there is an intimate relation, nay even an inseparable connection between Indra, Soma, V*ri*tra, the Dawn, the Sun, the Aryans, and Âryávarta or the Land of the Seven Rivers. Of these, A'ryávarta appears pre-eminently to be *the cradle of all*, at any rate *the primitive scene*; where, our R*ig*-Vedic ancestors and even their fore-fathers (पूर्वेभिऋषिभिः) had seen *Soma* grown on the Himâlayan Mountain, well watered and nourished by Parjanya or Rain, (as if in the capacity of Father) ; where only, and in no other region, the splendour and refulgence of the *Dawn* had first appeared to them on the horizon of the *Vipâsh*, *viz.* the modern river Beeas ; where, the brilliancy of the Dawn had subsequently vanished, and was absorbed in the dazzling light of the *Sun* ; where, the showers of rain were observed for the first time dropping from the clouds, represented as V*ri*tra, lying on the Indus, after these were rent asunder by Lightning represented by Indra's वज्र or thunderbolt, in the Land of the Seven Rivers ; where, our ancestors had performed sacrifices and continued their sacrificial sessions, which at times having been disturbed by outsiders, or even by our own kith and kin, such as the Zoroastrians, Indra was supplicated to give his aid to the Âryan progeny of the Land of the Seven Rivers, especially as he always defended his Ârya worshippers in all conflicts, nay, even punished those who were non-sacrificers and beyond the pale of our faith (इन्द्रःसमत्सुयजमानमार्यंप्रावत् ... R. V. I. 130. 8). Thus, the whole thing appears to have been most beautifully depicted in very lively colours, in the metaphorically terse language of the R*ig*-Veda.

All this, however, will certainly require an explanation and proof. I shall, therefore, hasten to furnish the same, as briefly as possible. There is evidence to show that Soma was born on, or that his origin has to be sought in, the mountain of the great Himâlayan chain (गिरिषु क्षयं दधे । R. V. IX. 82. 3), and that Parjanya was the father of Soma (पर्जन्यः पिता सोमस्य R. V. IX. 82. 3). For, Soma has been called Maujawata, simply for the reason that he was born on the Munjavat Mountain (सोमस्येव मौजवतस्य R. V. X. 34. 1) ; and as to the location of this mountain, we have the testimony of the Mahâ-Bhârata, from which we learn that the Munjavat Mountain lies on the slopes of the Himâlaya.

गिरेर्हिमवतः पृष्ठे मुंजवान्नाम पर्वतः ।

(महाभारते १४-८-१).

Thus, there appears no doubt whatever as to the origin of Soma having been in Aryâvarta, as he was born on the great Himâlaya Mountain. This Soma, however, being the Himâlayan Soma, has been considered from time immemorial to be of the best kind and of the finest quality. For, while commenting on (Rig-Veda X. 34. 1), Sâyana, the well-known Exegetist of our four Vedas, has strongly certified as to the superior quality of the Soma grown on the Munjavat mountain, and said, " मूजवति पर्वते जातो मौजवतः । तस्य । तत्रह्युत्तमः सोमो जायते ।" There is yet another place where Soma is grown, or at any rate used to grow previously, and the same appears to be the Sharyanâvat Lake, in the Province of Kurukshetra, as we find reference made to the fact in the R*i*g-Veda itself. For, says the verse, " Let V*ri*tra-slaying Indra drink Soma by Sharyanâvata's side." (Griffith).

शर्यणावति सोममिन्द्रः पिबतु वृत्रहा ।

(R*i*g-Veda. IX. 113. 1).

While making comments on this verse, Sâyana writes as under :—शर्यणावति । शर्यणावन्नाम कुरुक्षेत्रस्य जघनार्धे सरः । तत्रस्थितं सोमं वृत्रहेन्द्रः पिबतु ।

In the circumstances, there remains not the slightest doubt, in regard to the origin and growth of Soma, in Âryâvarta and Âryâvarta alone, and nowhere else ; and evidently, it was from here that Indra, after drinking Soma, carried his victorious arms to the North (उत्तरोभवन्), giving his ready assistance to the Sacrificers (R. V. I. 130. 8; ante *p* 40), and destroying the non-sacrificers (R. V. VIII. 14. 15).

असुन्वामिन्द्र संसदं विषूचीं व्यनाशयः । सोमपा उत्तरो भवन् ॥

(ऋ० वे० ८-१४-१५).

Turning to Indra, we find that Soma is his father (सोमः...जनितेन्द्रस्य R. V. IX. 96. 5), and that he (Indra) was born on the mountain in the very house of his father (पितुर्दमे R. V. III. 48. 2). We further perceive that Indra had asked for the *Soma-drink*, immediately after he was born, and that his mother had supplied him with it, even before suckling him (तं...माता परि...आसिंचदग्रे । R.V. III. 48. 2). All this, therefore, evidently proves the origin of Indra and Soma in Âryâvarta.

Now, as regards their age, it may be said that there is equally strong evidence in respect of the very great antiquity of both Soma and Indra. Speaking of Soma first, we find that the Rig-Vedic Rishis have declared him to be extremely old, saying that he existed even before the sacrifices (यज्ञस्य पूर्व्यः । R. V. IX. 2. 10). In fact, it seems that even *the ancestors* of our *hoary* Rig-Vedic fore-fathers had used Soma for sacrificial purposes (त्वया हि नः पितरः सोम पूर्वे कर्माणि चक्रुः पवमान धीराः । R. V. IX. 96. 11), and the more we look into his antiquity, the more he appears to be old (एष प्रत्नेन जन्मना देवो (नाम सोमः) Rig-Veda. IX. 3. 9). As to Indra, it may be mentioned in brief that, even in the Rig-Veda, it is said, while addressing this deity, that there is none superior to, and older than Indra (नाकिरिन्द्र त्वदुत्तरो न ज्यायाँ अस्ति वृत्रहन् । R. V. IV. 30. 1).

These being the very facts, we need not wonder if even the occidental scholars of note admit that Indra and Soma are the highest, the most ancient, and exclusively[1] the Hindu or

---

1 I may mention here, that the Zoroastrians were living with us—the Indo-A'ryans—in Aryávarta, before the religious schism. They, therefore, knew,

Indo-A'ryan Gods. In view, therefore, of bringing conviction home to the reader, I venture to give a few extracts from the works of some of them. Says Muir, " Indra is an exclusively Indian God." (*Original Sanskrit Texts.* vol. V. *p.* 119. Ed. 1870). Professor Roth, in his Lexicon, remarks under Indra that, " He was the favourite national deity of the A'ryan Indians in the Vedic age." Spiegel speaks about Indra as " the highest God of the earliest Hinduism. " (Introduction. Avesta). While Mr. Tilak says that, " the word Indra does not occur in European A'ryan languages " (The Arctic Home in the Vedas. *p.* 295), meaning thereby that even the word is not known in

---

when with us, Soma and Indra. Subsequent to the schism, and after they left Aryávarta, they settled in Iran. But, even then, they could not divest themselves of the primitive legends and mythologies they had imbibed and inherited from us, while living in Aryávarta, the land of their origin. Naturally, therefore, we find *Soma* transformed into *Homa*, and *Vritrahá*, which is an epithet of Indra, metamorphosed into Verethrajao, in Zend. Thus, we observe Dr. Windischmann remarking as follows, in his Dissertation " On the Soma-worship of the A'ryans." ......"But, by far the most remarkable analogy is that which exists between the Haoma of the Zendavesta and the Soma of the most ancient Brahmanical books,"......"*verethrajao*, Zend,—*vritrahá* Sanskrit (R. V. I. 91. 21), destroyer of enemies ".

Europe anywhere, much less the God of that name. Besides, even Dr. Taylor declares that, "In the Avesta there is barely a vestige of the great name of Indra, nor can it be traced in any of the European languages." (*The origin of the Aryas. p.* 312. Second Edition).

Now, in respect of Soma, Ragozin observes that, " From numerous indications scattered through the hymns (of the Rig-Veda), it appears probable that this (that is, the Himâlayan high-land of Kashmir) was the earliest seat of the Soma-worship known to the A'ryan Hindus, whence it may have spread geographically with the race itself." (Vedic India *p p.* 160-171. Ed. 1895). Moreover, from the description of Soma and its origin, given in the Mahâ-Bhárata and the Rig-Veda, (vide ante *p.* 41–42), it clearly appears that the Himâlayan high-land is *the only region favourable to its growth,* and that it grows or grew nowhere else. Naturally, therefore, it was not known in Europe, as Mr. Tilak writes to say that, " the word (Soma) is not found in the European languages." (The Arctic Home in the Vedas. *p.* 205).

Thus, we have tried to prove, and the evidence being irrefragable, have endeavoured to establish, that Indra and Soma having had their *origin* in A'ryávarta, retain an inseparable connection witn *the Land of the Seven Rivers*, where Indra with the aid of Soma had achieved the ever memorable exploits, which, to speak in the homely language, were nothing else than *the occurrence of the grand phenomena of Nature* of which, after our first progenitors—the ancestors of our Rig-Vedic fore-fathers—were born in A'ryávart, they were but naturally *the primitive observers*, in this land of their birth. Evidently, their observations had made *ever-lasting impressions* on their mind, and as they thought these were of great moment and of utmost importance, they had taken extreme care to *scrupulously preserve them intact*, and hand them down to *posterity* for their benefit.

Accordingly, it appears that our Rig-Vedic Rishis had but *traditionally* received *the first impressions* of our progenitors, and they had simply disburdened themselves by giving them an expression and a form of speech. They, thus, have bequeathed to us the richest legacy, the Mankind could receive, as it forms but *a true* and *a brilliant record* of our *first words, an authentic source* of our *first impressions*, and the *fountain-head* of our *genuine history, commencing even from our Cradle.*

I shall now advert to वृत्र, the Cloud-serpent, always found associated with Indra, and explain by whom, how, why, and where, he was killed. I shall then state the consequences that flowed from the action taken by Indra in respect of V*ri*tra, and mention the creations of Indra, as they happen to have had intimate connection with Aryâvarta.

We have already seen that Indra was very fond of *Soma-drink*, and that he having asked for it, on the very day he was born, his mother had supplied him with it (R*i*g-Veda. III. 48. 2 ; ante *p.* 43). We have further evidence to show that Indra had always received material aid from Soma, as it was only after enjoying the Soma-juice, that under its exhilaration, he was able to achieve the first and the great exploits (सुतस्य अस्य मदे अहिमिन्द्रो जघान । R. V. II. 15. 1 ; ...सोमस्य ता मद इन्द्रश्चकार । R. V. II. 15. 2 ; अन उषसः संपिपेष ! ...सोमस्य ता मद इन्द्रश्चकार ॥ R. V. II. 15. 6). And what are these exploits ? These are innumerable, and as such, a detailed description thereof would certainly fill up volumes. But, limited space at our disposal would not allow us to do that, and we shall only content ourselves by mentioning a few of them that are pertinent to our present investigations. Firstly, then , Indra (इन्द्रो), by means of his thunderbolt (वज्रेण), had killed V*ri*tra (अहन्वृत्रम् ...। R. V. I. 32. 5 ; अहन्नहिम् ...। R. V. I. 32. 1), *the first born of Serpents* (प्रथमजामहीनां... R. V. I. 32. 3), and as such *the oldest*[1] *and the greatest of them all* (वृत्रतरं R. V. I. 32. 5). Here, the expression *the first born of Serpents* and *the oldest and the greatest of them all* (प्रथमजामहीनाम्...वृत्रतरं) is really pregnant with meaning, and therefore will require an explanation, which I will presently offer. But, before I do that, I must remind the reader, that Indra's वज्र or thunderbolt represents *lightning*, while the clouds that hold water are represented as Serpents. These had held off *by force* (महिना R. V. I. 32. 8) the rain-waters, and thereby prevented them (याः पर्यतिष्ठत् R. V. I. 32. 8) from falling on *the Land of the Seven Rivers*. Indra, therefore, by means of his thunderbolt *cleaved the Serpents* (भिन्नं [वृत्रं]) and rent asunder the clouds. The waters then forth-with trickled down from the clouds (स्यन्दमाना अंजः), and flowed to the sea. (समुद्रमवजग्मुरापः। R. V. I. 32. 2). Thus, the metaphor is complete, the meaning is clear, and it is all plain sailing, so far. We shall, therefore, turn our attention for a while to the expression "*the first born of Serpents*", used in regard to Vritra, as it has a very deep meaning. Nay, it has also vast importance and *significance*, *all its own*. I shall therefore endeavour to explain this as briefly as possible.

---

1 As between Indra and Vritra, the former is still the more ancient. (Vide ante p 43. R.V.IV. 30. 1).

We have already seen that अहि or *Serpent* means मेघ or *cloud*, and Sâyana explains the word *Ahi* as meaning *Megha* (अहीनां = मेघानां। R. V. I. 32. 3). In like manner, V*ri*tra also means serpent; while, Indra is designated as the slayer of V*ri*tra, who withholds the waters in the clouds and the necessary rain. Evidently, destruction of V*ri*tra involved the fall of rain, and therefore, he was killed by Indra, and the waters in the clouds let off. We also know that the fall of *Rain*, the rise of the *Sun*, and the splendour of the *Dawn* before sunrise, are but the phenomena of Nature, which must have certainly occurred before the advent of Man ; nay, these perhaps used to occur even from the beginning of things.

But, all this notwithstanding, our *Primitive fore-fathers*—the oldest ancestors of our R*i*g-Vedic bards—could but mark, very naturally enough, these phenomena, when *only* they were able *to stand* on their own legs, or were able to *see* and to *observe*, or *to appreciate* and *to admire*. It was, therefore, *then only*, that they thought they had *for the first time observed waters* dropping from *the serpent-like-clouds* which had spread over the endless expanse of space hovering over their heads in the sky; which had held off waters within them; and which were only let off on the earth, after pierced through and smitten [भिन्नं

R. V. I. 32. 8] by lightening, produced by the all pervading *Power*, then designated as Indra. These clouds, therefore, were naturally thought to be *the first born*, *the oldest*, and *the greatest* (प्रथमजामहीनाम् R. V. I. 32. 3 ; वृत्रं वृत्रतरं R. V. I. 32. 5). In fact the waters that dropped from the clouds, were supposed to be *the first showers*, *the first rain-fall*, and *the first Serpent* killed by Indra ; because, they had come within the *observation* of our *Primitive* ancestors, *only for the first time*.

Thus, the *first observation* by our *primitive ancestors*, in respect of the fall of rain from the clouds having been established, the very legitimate question would arise, where had these phenomena occurred ? Where were the rain-clouds seen ? And where had the rain-waters dropped down ? This, therefore, we shall proceed to prove presently. From R*i*g-Veda (I. 32. 11), we perceive that Indra had, by killing V*ri*tra (वृत्रं जघन्वान्), thrown open (अपववार) the gates of waters (अपां बिलं), which were kept closed (अपिहितं यदासीत्) by V*ri*tra-serpent, who, moreover, had held off the waters, which had stayed in the clouds (अतिष्ठन्), as if confined (अहिगोपा निरुद्धा आपः) like kine held by a robber (पणिनेव गावः). And now the very next verse of the R*i*g-Veda answers the question and informs us, that the waters held off, or imprisoned in the cave (बिलं) by *Vritra*, and subsequently let off (अवासृजः), or restored to freedom by Indra after conquest (अजयो गाः), had flooded *the Seven* Rivers, as it was intended by God that *these should flow* (सर्तवे सप्तसिन्धून्). This, in other words, means that after thunder and lightning, there were heavy showers of rain from the clouds, which having filled the सप्तसिन्धवः or the *Seven Rivers of Âryâvarta*, they began to flow. (...इन्द्र...अजयोगाः...शूर... अवासृजः सर्तवे सप्तसिन्धून्। R. V. I. 32. 12). Here, the words *Sapta-Sindhûn* certainly refer to *the renowned Seven Rivers* of Âryâvarta, and the previous hoary traditions support this view. For, says Sâyana as follows, while commenting on R*ig*-Veda (I. 32.12): सप्तसिन्धून्। इमं मे गंग इत्यस्यामृच्याम्नाता गंगाद्याः सप्तसंख्याका नदीः सर्तवे सर्तुं प्रवाहरूपेणगन्तुमवासृजः।

But, more than this, and as if to corroborate the fact and fortify our conclusions, we have further testimony to prove that the clouds from which waters were *first observed* by our ancestors, trickling down the Earth and flooding the Seven Rivers, were hovering over the great Indus of Âryávarta: and the *Vritra*-Serpent that had withheld waters in the clouds and was therefore killed by Indra, was also lying concealed in the clouds, that were hovering over this very Indus. The evidence in this respect is of the utmost importance, and we make no apology to give it here in full. For, in (II. 11. 9 of the Rig-Veda), the poet says that, " Indra (इन्द्रेा) hurled down and killed (अवहन्निः) deceitful Vritra (मायाविनं वृत्रं), that was lying concealed (आशयानं) on the great Indus (महासिंधुम्). Here, the words used in the verse are not only *plain* and *simple*, but *unequivocal* and *distinct*, and as such convey a very clear meaning ; since, महा means the great, and सिंधु means the Indus, one of the Seven renowned Rivers of Âryávarta, which Indra had let flow, after having slain the Vritra-Serpent (यो हत्वाहिमरिणात्सप्तसिन्धून् R. V. II. 12. 3).

Now, after this phenomenon, viz. *the first observation*, by the Primitive ancestors of our Rig-Vedic fore-fathers, *of the fall of rain in the Land of the Seven Rivers*, or to speak in the usual Vedic metaphor, *the observation* of the destruction of Vritra-Serpent, which appears to have been the *first* heroic deed, out of the several, achieved by Indra at the very commencement (इन्द्रस्य नु वीर्याणि प्रवोचं यानि चकार प्रथमानि वज्री । अहन्नहिं... ॥ ऋ०वे० १-३२-१), we come next to Indra's second and the third deed of heroism, which, in short, consists of the creation of the Dawn and the Sun (आत्सूर्यं जनयन्द्यामुषासं...। R. V. I. 32. 4). These the Rig-Vedic Poets beautifully portray in very glowing colours, and while giving a graphic description of them, when their wonderful splendour was perceived above the horizon, they but inadvertently throw a surprisingly agreeable hint in respect of the region where these were *first observed by our Primitive ancestors.* This careful *observation*, therefore, affords but another *clue to our Âryan cradle* and *Original Home*, and as it seems to be of great moment, I venture to give the original text *in extenso*, along with its translation, as rendered into English by Oriental and Occidental scholars.

We have already seen that the first deed of heroism achieved by Indra was the slaying of V*ri*tra-Serpent and the consequent fall of rain—a phenomenon of Nature—and that his next exploits, among others, were the creation of the Dawn and the Sun (यदिन्द्राहन्प्रथमजामहीनाम्... , आत्सूर्यं जनयन्द्यामुषासं... R. V. I. 32. 4). Now, the chief thing, with which we are here concerned, in regard to the fact after their creation is : *Where was the Dawn first seen ?* And *Where* was the *Sun* first observed by our *Primitive Ancestors ?* We all know that the delightful *Dawn*, the splendour of the morning *Sun*, and the Dawn disappearing in, or absorbed by, the rising Sun, are but the usual phenomena of Nature. And the only question is, where had these occurred, or observed by our Primitive ancestors ? We must, therefore, go to the R*i*g-Veda, and seek its assistance for answering the queries. In the R*i*g-

Veda (IV. 30. 8; IV. 30. 9; and many other places), we find उषस् or the Dawn represented as the daughter of the Sky or Dyaus (दिवः... दुहितरं... उषासम्... R. V. IV. 30. 9) and Indra (meaning the Sun, as in R. V. I. 6. 3; 1. 26. 1; IV. 26. 1; X. 89. 2;) described as having crushed her (उषासामिन्द्र संपिणक्... । R. V. IV. 30. 9). After this, however, another verse states to say that, Ushas had fled away affrighted (अपोषा ...सरत्... बिभ्युषी), because her car was smashed by Indra (अनसः... संपिष्टात्... नि यत्सीं शिश्नथद्वृषा । R. V. IV. 30. 10). This, evidently, is the morning phenomenon, and naturally conveys the idea that the splendour of the Dawn or the morning light, having first appeared, had disappeared after a while, in the bright rays of the Sun, or rather was absorbed in it, after its appearance on the horizon.

But, the essential point has yet remained unsolved, and, as such, we must endeavour to find out the place or the region, where the Dawn (उषस) had disappeared in the glowing light of the Sun. Fortunately for us, after making requisite researches in the inexhaustible mine of the Rig-Veda, we come across the thing sought for. Because, the Rishi, Wâma Deva, informs us to say that, "the Dawn was crushed by Indra, as she appeared proudly (महीयमानाम् । उषासमिन्द्र संपिणक् ॥ R. V. IV. 30. 9); that her car lay smashed on the river Vipásh, after it was broken by Indra; and that she had afterwards fled away from thence." The original is as under:—

> एतदस्या अनः शये सुसंपिष्टं विपाश्या ।
> ससार सीं परावतः ॥ (ऋ० वे० ४. ३०. ११).

" That car of hers lay smashed on the Vipat. Away she ran from thence." (S. P. Pandit).

" So, there, this car of Ushas lay, broken to pieces, in Vipâs. And she herself fled far away."
(R. T. H. Griffith).

It is hardly necessary here to state, that the Dawn (उषासम्) and the Sun (सूर्य) were created in Âryâvarta by Indra, who was also of Âryâvarta, as the Dawn had appeared and disappeared in the region of the river Vipâsh (modern Beeas), and the Sun also had made its appearance on the horizon of the same rigion, viz. Beeas, which evidently forms part of the Punjâb. (vide Rig-Veda. IV. 30. 9, 11).

Here, then, we stand on solid ground and *terrestrial evidence*. For, the river Vipâsh[1] referred to in the aforesaid verse is the modern Beeas,[1] and Yâska identifies it with the river

1 Vide Max Muller's *India. What can it teach us?* *p.* 172. Edition 1883 ; and Muir's *Original Sanskrit Texts.* vol. II. *pp.* 345. 342, Note 116. Edition 1871.

Ârjîkîyâ, as he says:— (महाकूलार्जीकीयां विपाळित्याहुः .... । निरुक्ते उ० प० अ० ३. २६). This, therefore, is not an imaginary river, but a real terrestrial river of the Punjâb, as Max Muller distinctly says that, "It was probably on the Vipâsh (later Vipâshâ), a north-western tributary of the Sutledge, that Alexander's army turned back. The river was then called Hyphasis ; Pliny calls it Hypasis, a very fair approximation to the Vedic Vipâsh, which means 'unfettered'. Its modern name is "Bias or Bejah." (What Can India Teach us? *p.* 172. Edition 1883). Moreover, if we bear in mind the celebrated and the oft-quoted verses (इमं मे गंगे यमुने सरस्वति... ।।५।। तृष्टामया....। त्वं सिंधो कुभया गोमतीं क्रुमुं मेहत्न्वा सरथं याभिरीयसे ।।६।। R. V. X.75), it will certainly be easily perceived that the Poets of the R*i*g-Veda-period had, by all means, a wider geographical horizon than has usually been supposed, and had, therefore, very accurate knowledge of *the Land of the Seven Rivers.*

Thus, to come nearer home, it seems evident that the Dawn (उषस्) also, was, *like the rain-waters, first observed* in her dazzling splendour in the region of the river Vipâsh–the modern Beeas–by our *primitive ancestors,* who had also seen her, for the *first time disappearing* and absorbed in the unsurpassed refulgence of the Sun that was coming on the horizon, in the region of that very river, flowing in Âryâvarta. Now, if *the Primitive Fore-fathers* of our Rig-Vedic ancestors had *seen the first drops of water*, or *the showers of rain*, *dropping from the clouds, before every thing else, in the Land of the Seven Rivers*, and nowhere in Central Asia, nor in Northern Europe, nor in the Arctic Regions; if they were the *first observers* of *the light* or *the Dawn, in the region of Beeas* (the Vipâshâ); and if, moreover, they had also *seen the Sun for the first time*, rising above the horizon in *the region* of this very river—the Vipásh—of Âryâvarta, and nowhere else; then, certainly, the natural inference is, that *the Land of birth of our Primitive Ancestors, was Âryávartá* and Âryávarta only, and *no other whatever*. As such, therefore, *we are autochthonous in India*, and *not immigrants into the Land of the Seven Rivers*.

Moreover, we find that Indra, the Lord of the whole Universe (विश्वस्य जगतः प्राणतस्पतिः। Rig-Veda. I. 101. 5), had, *first* and *before all*, (प्रथमो) procured *water*[1] and *light*[1] (गा अविन्दत्)

1 गाः (धेनवः) or Cows mean, in the Rig-Veda, both Light and Water. (vide R.V. I. 32. 11 ; I. 92. 1 ; III. 1. 7 ; IV. 22. 6). For, while in (I. 92. 1) Sâyana interprets गाः as meaning रश्मीन् (rays), in (IV. 22. 6) he explains धेनवः सिस्रते by saying गावः प्रकर्षेण क्षीरं क्षरंति। "The rain makes showers of fertilising water."

in the region of the *Seven Rivers* (ante *pp.* 51-57), *for his worshippers*—the *Bráhmans*[1] – (ब्रह्मणे, R. V. I. 105. 5), who were *pre-eminently* the *Sacrifice-loving Âryâns*, our *Primitive ancestors*. As the verse is very important, I herein below give it in full, with Griffith's translation :—

यो विश्वस्य जगतः प्राणतस्पतिर्यो ब्रह्मणे प्रथमो गा अविन्दत् ।
इन्द्रो यो दस्यूँरधरां अवातिरन्मरुत्वन्तं सख्याय हवामहे ॥ ५ ॥

(Rig-Veda. I. 101. 5).

" He, who is the Lord of all the world that moves and breathes, who for the Brâhman, first before all, found the Cows ;

" Indra who cast the Dasyus down beneath his feet,— him girt by Maruts we invoke to be our friend."

Here, *Dasyus* are evidently the V*ritra*-Serpents,[2] who having held off waters in the clouds, were hurled down on the Earth, and crushed beneath his feet by Indra (इन्द्रो... दस्यूँ

---

1 The word Brâhmaṇa means one who knows *Brahma* or *the laud of the Lord* ; and Sâyana, the great exegetist, while commenting on (R. V. I. 47. 2.), defines *Brahma* as follows :— "ब्रह्म स्तोत्ररूपं मंत्रं.... (कुर्वन्ति). " Bráhmaṇa therefore is one who knows the *laud*, that is *Brahma* or Mantra:— ब्रह्म जानाति ब्राह्मणः ।.

2 Sâyana also interprets *Dasyú* as V*ritra*. For, in commenting on (R. V. I. 33. 4), he says :— "दस्युं चोरं वृत्रं...." ।.

अधरानवातिरत्), and *let the waters flow for* the *Brâhman worshippers* (ब्रह्मणे), for *the first time, and before all* (प्रथमो).

But, more than this, there is yet a further distinct statement that the rain was created and the showers were made (अददाम्... अहं वृष्टिं R. V. IV. 26. 2) only for the sake of the *Sacrifice-loving mortals* (दाशुषे मर्त्याय), and these certainly were *our Primitive Âryáns*, the ancestors of our Rig-Vedic Fore-fathers. And above all, as if to signify his special favour to the *Sacrifice-loving* Âryans—*our Primitive ancestors*—, and to make his own seal in token of approval, Indra had also *given land* (भूमिमददाम्) *to our Primitive Ancestors* (आर्याय).

अहं भूमिमददामार्यायाहं वृष्टिं दाशुषे मर्त्याय।

(Rig-Veda. IV. 26. 2).

" I have bestowed the earth upon the Ârya, and rain upon the man who brings oblation."

And what land or region could this be, other than the one, where in company with, and by the assistance of Soma, Indra (who like Soma, the Reader should remember, was born in Âryâvarta, vide ante *pp.* 41-42) had made the waters of the renowned Seven Rivers of Âryâvarta flow for the sake of our *Progenitor* Manu (इन्द्रो अपो मनवे सस्रुतस्कः।... अरिणात् सप्तसिन्धून्... R. V. IV. 28. 1), after slaying the Vritra-Serpent

(अहन्नहिं) that lay concealed on the Indus (महासिंधुमाशयानं II. 11. 9), and opening the channels (अपावृणोत् ... खानि), that were as though altogether shut (अपिहितेव ... Rig-Veda. IV. 28. 1)?

Besides, we next find that like Indra, the Sarasvatî too, *the most sacred* river of Âryâvarta (vide ante *pp*. 35,38), had killed the V*ri*tra-Serpent that was born and destroyed also in the region of the Indus (सिंधुं) *p*. 52. For, she seems to have obtained even the epithet वृत्रघ्नी on that very account (सरस्वती घोरा... । वृत्रघ्नी... ॥ ऋ० वे० ६-६१-७), and had made the waters flow for the man (विषमेभ्यो अस्रवः R. V. VI. 61. 3). Nay, she had even *procured* for and given land to him (उत-क्षितिभ्योऽवनीरविन्दः R. V. VI. 61. 3), and had made us (the sacrifice-loving Âryans) spread ourselves beyond all foes and beyond the regions of her other sisters (सानोविश्वा अतिद्विषः स्वसॄरन्या ऋतावरी । अतन्...॥ R. V. VI. 61. 9), viz., the renowned *Seven Sister-Rivers* of Âryâvarta (सप्तस्वसा सरस्वती R. V. VI. 61. 10), indicating thereby, that she had given opportunity and every sort of encouragement to our *Primitive Aryan ancestors*, and to our Rig-Vedic forefathers, to carry their victorious arms beyond the confines of the *renowned Seven Rivers*, and to extend their Colonies beyond Âryâvarta, in all directions.

The cumulative evidence, therefore, obviously supports the view that it was *the Land of the Seven Rivers* that was the *Cradle* of the *Primitive A'ryans—our Progenitors.* And, naturally enough, it was for this reason that even Professor Spiegel in his Introduction to Avesta, vol. II. *p.* CX, seems to have been inclined to entertain the view that " An origin in that locality (viz. the Country bordering on the Indus) might, with most probability, be ascribed to the legend of V*ri*trahan, as Indra is designated, as the slayer of V*ri*tra, who withholds the clouds and the necessary rains."

Now, our progenitors were certainly a *very gifted race* on the surface of the Earth ; while, our R*i*g-Vedic ancestors, having been their direct descendants, had inherited all the best traits of character and their finest qualities, had welcomed their hoary traditions and pure usages, and respected the sacrificial rites and religious ceremonies. They had first remained in *the Land of the Seven Rivers*–their *Home* and *Cradle*–deeply engaged in their sacrifices and varied rituals ; till, distant climes having excited their curiosity, it kindled their thirst for fame, and induced them to spread the A'ryan civilization beyond the Land of the Seven Rivers in all directions, to extend their Colonies everywhere, and to carry their victorious arms, all over the globe. Indra, accordingly appears, in right earnest, to have been supplicated as under, in the following verse:

अप प्राच इन्द्र विश्वाँ अमित्रानपापाचो अभिभूते नुदस्व ।
अपोदीचो अप शूराधरा च उरौ यथा तव शर्मन्मदेम ॥ १ ॥

(R*i*g-Veda. X. 131. 1).

" Drive all our enemies away, O Indra, the western, mighty Conqueror, and the eastern.

Hero, drive off our northern foes and southern, that we in thy wide shelter may be joyful." (Griffith).

Thus, impelled by a new spirit of adventure and conquests abroad, our ancestors of yore had left their Mother-land-A ryâvarta and crossed the Hindu-Kusha, and after traversing the vast Asiatic Plateaux had gone even as far as the A'rctic regions; where they had brought under their sway extensive tracts, which having been colonised by them, they had remained there for a considerable time, enjoying the genial climate and pleasant long days, expressing their astonishment at the wonderful sight of continuous Dawns lasting for days and months together, and also manifesting their overwhelming dread of the supposed endless nights, keeping, however, all the while, constant communication with their Mother-Country Aryâvarta, till at the advent of the Great Ice Age, compelled, all of a sudden,—by devastating floods of Ice, affecting the higher latitudes, alike of the Old and the New-World,—to return to their Original Home in A'ryávarta, or settle in countries which could give them refuge. And we have irrefragable evidence in support of these facts, which, therefore, I proceed to place before the Reader presently.

It has been scientifically proved, and al Geologists have unanimously agreed, that the climate of the A'rctic Regions was milder[1] before the advent of the Great Ice Age or the Pleistocene Period, and that owing to the existence of genial climate there, in the Tertiary Epoch, these regions had become habitable. It was evidently then, that our older ancestors had, after leaving Âryávarta, colonised them and settled there. This has been proved by very ancient sources of evidence of the hoary past; and the

1 For, Dr. Dana the American geologist says, "The mild temperature of the A'rctic is further evident from the.... formation of thick strata of lime-stone (which) shows that life like that of lower latitudes not only existed there, but flourished in tropical profusion." (Manual of Geology. *pp*. 224-5. Ed. 1863).

Rig-Veda, the pre-eminently oldest[1] of them all, affords ample testimony in regard to the facts alleged.

For, the Rishis of the Rig-Vedic period speak of the ever-lasting Dawns, long days, long nights, and even of six monthly day and six monthly night of the year, as if they had *personally observed these phenomena* and *witnessed the incidents as they had occurred*. We, shall, therefore, examine and analyse a few of them, for arriving at the right conclusion. In the Rig-Veda (V. 79. 9), the Dawn (दुहितर्दिवो) is asked not to

1 In regard to this, Professor Max Muller says that, "The principal charm of Vedic antiquities" is "its independent originality", (Rig-Veda. vol. IV. *p.* XXXVIII, 1862), and declares that the Rig-Vedas are "The spring-heads of the thought, of the language, and of the Poetry of India, which rise from depths inaccessible to foreign tributaries, and whose earliest course we may follow step by step in the literature of the Bráhmaṇas with greater accuracy than is the case in the early history of any other nation.

(Rig-Veda. vol. IV. *p.* LXXI).

He also gives emphatic expression to the thought that "The Rig-Veda is the most ancient-book of the A'ryan world", and "the sacred hymns of the Bráhmans stand unparalleled in the literature of the whole world, and their preservation might well be called miraculous."

(Rig-Veda. vol. IV. *p.* LXXX).

delay or tarry long (मा चिरं तनुथा), indicating thereby a very strong desire on the part of our ancestors to see the Sun rise, coupled with a wish that the Dawn should no more linger long on the horizon. In another place, there is a clear expression of feeling that a period of *several days* has elapsed between the first appearance of the Dawn on the horizon and the actual rising of the Sun which followed it (तानीदहानि बहुलान्यासन् या प्राचीनमुदिता सूर्यस्य । R*i*g–Veda. VII. 76. 3). But to crown the whole, we see our Vedic R*i*shis calling the Dawn everlasting, and saying that " the Goddess (उषस्) Ushas dawned continually or perpetually in former days " (शश्वत् पुरोषा व्युवास देवी । R. V. I. 113. 13). And as the Dawns appeared every time they were seen, new and young again, with shining splendour and surpassing brilliancy (नव्या नव्या युवतयो भवन्तीः), for days and months together continuously, the phenomenon having been new, and such as was never observed by our ancestors of old before, while they were in their Mother-land A'ryâvarta, it naturally created astonishment in their minds, and in wonder said, "*the great divinity of the Gods is unique*" (महद्देवानामसुरत्वमेकम् ॥ Rig-Veda. III. 55. 16).

We also find indisputable evidence in the Rig-Veda of long days and nights, as one passage therein expressly says that, "the Sun extended his daily course to an unusual length (दीर्घं ततान सूर्यो न योजनम्। ५-५४-५; Rig-Veda. V. 54-5). In another place, there is still stronger testimony, as it clearly indicates that, "*the Sun unyoked his car in the midst of heaven*, and the A'rya, (that is Indra, the friend, supporter, and the God of Gods of the Indo-A'ryans), found a counter-measure (प्रतिमानम्) for the Dâsa," the enemy of the Âryas and the creator of darkness. As the verse is very important or our purpose, I quote it here below :—

वि सूर्यो मध्ये अमुचद्रथं विददासाय प्रतिमानमार्यः।

(R. V. X. 138.3).

Griffith renders this into English as follows :—"In the midway of heaven, the Sun unyoked his car: the A'rya found a match to meet his Dâsa foe ".

There are also verses in the Rig-Veda which unmistakably indicate long, very long, nay too long nights, that lasted for days and months together. For, in (I. 46.6), the Ashwins seem to be requested to vouchsafe such strength to the poet and the worshipper, as may carry him through the darkness. (या नः पीपरदश्विना ज्योतिष्मती तमस्तिरः।). While in (II. 27.14), the poet says, "May I obtain the wide fearless light, oh Indra! May not the long darkness come over us." (उर्वश्यामभयं ज्योतिरिन्द्र मा नो दीर्घा अभिनशन्तमिस्राः ॥). And in (X. 124.1), Agni is told that he had stayed very long (दीर्घं), nay too long (ज्योक्), in darkness (ज्योगेव दीर्घं तम आशयिष्ठाः ।). Moreover, in another place, we have an express statement in which Night is invoked "to become fordable (सुतरा), with ease and without any difficulty " (अथा नः सुतरा भव ॥ X. 127.6). But, more than this, we have in the Atharva Veda, a very distinct expression and clear state of mind of our older, nay *Tertiary ancestors*, in respect of the extreme length of night and the tiresome darkness, which they were actually afraid of, having not been accustomed to it before while in A'ryâvarta, as they seem to have exclaimed in great disappointment and dismay that, " its (Night's) yonder boundary is not seen " (न यस्याः पारं ददृशे । A. V. XIX. 47.2). While, in the Taittirîya Samhitâ, there is a statement which clearly says, after addressing the Night, " O Chitrâvasu ! Let me safely reach thy end " (चित्रावसो स्वस्ति ते पारमशीय । T. S. I. 5.5.4). It then explains the very incident and the cause of it tersely, as follows :-" Chitrâvasu is (*i. e.* means) the night, and as it was apprehended, in the by-gone age, that the night would not come to an end, the

Brahmans were afraid that it would not dawn." (रात्रिर्वै चित्रावसुरव्युष्ट्यै वा एतस्यै पुरा ब्राह्मणा अभैषुः । Taitt. Sam. I. 5.7.5).

And this reminds us of the penance or the Prâyaschitta (प्रायश्चित्त) which the Gods had to perform, whenever the Sun did not rise or shine as was expected. For, the Taittirîya Samhitâ supplies us with the requisite evidence on the point, as it says that, the Devas (Gods) had to perform the Prâyaschitta on the occasion.

असावादित्यो न व्यरोचत तस्मै देवाः प्रायश्चित्तिमैच्छन् ।

(Taitt. Sam. II. 1 2. 4).

Thus, while on the one hand, we find our Vedic ancestors—after they left A'ryâvarta, their Mother-Country, and reached the A'rctic regions which they subsequently colonized—delighted to see the everlasting or rather continuous Dawns, shining for days and months together, with increasing brilliancy, followed by long days, with expressions of wonder at the unusual sight (महद्देवानामसुरत्वमेकम्), having not been accustomed to it during their stay in A'ryâvarta—*their Home and Cradle*—we, on the other hand, observe our Tertiary forefathers, exhibiting extreme impatience for *the end of nights* and for the morning light, entertaining misgivings about the coming Dawn, nay even manifesting their *utmost fear* that *darkness would never come to an end* (रात्रिं रात्रिमतरिष्यन्तस्तरेम तन्वा वयम् । न यस्याः पारं ददृशे । A. V. XIX. 47.2), and expressing desire that they may pass uninjured in their body, through each succeeding night.

The expression of amazement and wonder (महद्देवानामसुरत्वमेकम् R. V. III. 55), at the sight of an unusually long day and night lasting for months together, was evidently subsequent to our leaving A'ryâvarta (आर्यावर्त) —our original Home—and after we reached and colonized the A'rctic Regions. For, while in A'ryâvarta, we were not at all accustomed to the phenomena of long days and long nights. But, when, after going to the A'rctic regions from A'ryâvarta, our Tertiary ancestors had noticed them all of a sudden, it naturally caused amazement at the sight of long[1] Dawns (R. V. I. 113. 10 ; III.55.16) and long[2] days (R. V. X. 138. 3), and struck terror into the hearts of our hoary ancestors when visited by tiresome nights, as they could not see the end of darkness and of

---

1 कियात्या (उषसः) यत्समया भवाति. ॥(ऋ० वे० १-११३-१०).
नव्या नव्या युवतयो भवंतीर्महद्देवानामसुरत्वमेकम् ॥
(ऋ० वे० ३-५५-१६).

2 वि सूर्यो मध्ये अमुचद्रथं दिवो विददासाय प्रतिमानमार्यः
(ऋ० वे० १०-१३८-३).

the long[1] nights (R. V. I. 32.10; II. 27.14; X. 127.6; A. V. XIX. 47.2).

It may not be out of place here to state, that *in the A'rctic regions*, the continuous long nights without a break, lasted for so long a period as even six months, like six months' long days, thus making a year consist of *one long night* and *one long day*, each of six months' duration, and thereby confirming the tradition found in the Taittirîya Brâhma*n*a which says, "That which is a year is but a single day of the Gods".

एकं वा एतद्देवानामहः यत्संवत्सरः। (Taitt. Br. III. 9.22.1).

Moreover, in the Taittirîya Âra*n*yaka and in the R*i*g-Veda, we find the year personified, and in the text of the former, there appears a distinct expression that to the right and the left side of the Year-God, there are the bright and the dark days. The following verses refer to this, and the same being important for our purpose, I quote them here below:—

शुक्लकृष्णे संवत्सरस्य। दक्षिणवामयोः पार्श्वयोः। तस्यैषा भवति (T. A. 1-2-4).

---

1 दीर्घं तम आशयदिन्द्रशत्रुः......(ऋ० वे० १-३२-१०.)
मा नो दीर्घा अभिनशन्तमिस्राः ॥ (ऋ० वे० २-२७-१४).
अथा नः सुतरा भव ...... ॥ (ऋ० वे० १०-१२७-६).
न यस्याः पारं ददृशे ......(अथर्ववेद १९-४७-२).

शुक्रं ते अन्यद्यजतं ते अन्यद्विषुरूपे अहनी द्यौरिवासि । (R. V. VI. 58.1).

This means in substance, that the Year-God's one form is bright while the other is sacrificial or dark. But, in another place, the Taittirîya Âra*n*yaka goes still further and says that, "the year has one head and two different mouths," observing at the same time that, "all this, is but the season-characteristic."

एकं हि शिरो नाना मुखे। कृत्स्नं तदृतुलक्षणम्। (Taitt. Ara*n*. 1.2.3).

And the tradition of one whole year having been divided into *Ahanî* (*nycthemeron*), consisting of one continuous day and night, each of six months' duration, which in fact was actually witnessed by our R*i*shis of the R*i*g-Veda period, appears to have been preserved not only in the later Vedic works, but also in the Mahâ-Bhârata, the मनुस्मृति or the Code of Manu, and the astronomical works like the Sûrya-Siddhânta :—

बभूव रात्रिर्दिवसश्च तेषां
संवत्सरेणैव समानरूपः ॥ १३ ॥ (Mahâ-Bhârata. III. 165). South Indian Texts Edition. 1908.

दैवे रात्र्यहनी वर्षं प्रविभागस्तयोः पुनः
अहस्तत्रोदगयनं रात्रिः स्याद्दक्षिणायनम् । (Manu Smr*i*ti. I. 67).

मेरौ मेषादिचक्रार्द्धे देवाः पश्यंति भास्करम् ।

सकृद्देवोदितं तद्वद्सुराश्च तुलादिगम् ॥ (Sûrya Siddhânta. XII. 67).

Now, the phenomena of ever-lasting or rather continuous Dawns, long days, and long nights, are the sure characteristics of the Polar regions, and as these traditions are always "founded on the observation of Nature," as remarked by Sir Charles Lyell in his Elements of Geology (vol. I. *p.* 8. Edn. 11th), we have every reason to suppose that at one time, in the bygone-Age (युग), our R*i*g-Vedic ancestors and their forefathers had certainly colonised the A'rctic regions and made vast settlements there. It appears, therefore, necessary to find out when this was done, and also to see whether there is any evidence to fortify our conclusions.

It seems, that the climate of the A'rctic Circle was genial, and the regions, as far as the North-Pole, were habitable in the Tertiary Epoch and before the place was invaded by Ice, which not only occupied the higher latitudes but rendered the A'rctic regions unsuited for the habitation of man, beast, and even plant. In the light of these geological facts, therefore, as also Vedic evidence, placed before the reader heretofore, it appears that in the Tertiary Epoch and before the advent of the Great Ice Age, the *adventurous progeny of our Primitive fore-fathers* had, after leaving their Mother-Country-*A'ryâvarta*,–

colonised the Arctic regions and made settlements almost everywhere, as the climate was mild and genial at the time, keeping all the while, constant, nay uninterrupted communication with their Mother-land A'ryâvarta, which they loved most.[1] In the Arctic colonies, they

---

1 Our older ancestors seem to have had an ideal love for their Mother-Country—A'ryávarta— the renowned Land of the Seven Rivers, and the *pathos* with which they sing her praises and the valour of their A'ryan Race, nay speak of the glory and Majesty of her most stupendous Himalayan Mountains, describe the purity of the most sacred river the Sarasvati, and singularly mark out her lovely Regions, appears simply unique. While, the unbounded patriotism of our ancestors appears to run through every vein, and seems to be beyond all comparison. I, therefore, venture to give a few extracts from the ancient Sanskrit Literature, as I think they will certainly repay perusal :—

(1) अजयो गा अजयः शूर सोममवासृजः सर्तवे सप्तसिन्धून्॥
(R. V. I. 32. 12).

उत नः प्रिया प्रियासु सप्तस्वसा ......सरस्वती ॥ १० ॥
(R. V. VI. 61. 10).

विजानीह्यार्यान्ये च दस्यवो बर्हिष्मते रन्धय शासदव्रतान्
(R. V. I. 51-8).

(2) पावका नः सरस्वती । (Rig. Veda. I. 3-10).

त्वे विश्वा सरस्वति श्रितायूंषि देव्याम् ॥
(Rig. Veda. II. 41. 17).

(3) नानावीर्या ओषधीर्या बिभर्ति......॥
(Ath. Veda. XII. 2).

had remained for a considerable time, till forced to leave these once genial tracts, owing to the overpowering floods of Ice, and take resort to the then, as now, the highest Mountains in the

---

(4) यस्यां समुद्र उत सिंधुरापो यस्यामन्नं कृष्टयः संबभूवुः ।
(A. V. XII. 3).

(5) यस्यां पूर्वे पूर्वजना विचक्रिरे यस्यां देवा असुरानभ्यवर्तयन् ।
(A. V. XII. 5).

(6) गिरयस्ते पर्वता हिमवन्तोऽरण्यं ते पृथिवि स्योनमस्तु
(A. V. XII. 11).

(7) भूम्यां देवेभ्यो ददति यज्ञं हव्यमरंकृतम् । (अ. वे. १२ १.२२).
यस्यां गायन्ति नृत्यन्ति भूम्यां मर्त्या व्यैलबाः ।
युद्धं ते यस्यामाक्रंदो यस्यां वदति दुन्दुभिः ॥
(A. V. XII. 1. 41).

ऋषयो वै सरस्वत्यां सत्रमासत । (Ait. Br. 2. 19).
माध्यमाः सरस्वत्यां सत्रमासत ।
(Koush. Br. 12. 3).

(8) ग्रीष्मस्ते भूमे वर्षाणि शरद्धेमन्तः शिशिरो वसन्तः ।
ऋतवस्ते विहिता हायनीरहोरात्रे पृथिवि नो दुहाताम् ॥
(A. V. XII. 36).

This description of six seasons and of equal days and nights, eminently suits A'ryávarta.

(9) सरस्वती दृषद्वत्योर्देवनद्योर्यदंतरम् ।
तं देव निर्मित देशं ब्रह्मावर्तं प्रचक्षते ॥
(Manu. Smr. 2. 17).

अत्रापि भारतं श्रेष्ठं जंबूद्वीपे महामुने
यतो हि कर्मभूरेषा ततोऽन्या भोगभूमयः ॥ २२ ॥

अत्र जन्मसहस्राणां सहस्रैरपि सत्तम ।
कदाचिल्लभते जन्तुर्मानुष्यं पुण्यसंचयात् ॥ २३ ॥

world, the Himâlayas, for returning to A'ryâvarta *their original Home and Cradle.*

And the fact, that when after the Great Deluge, caused by the incessant and heavy Ice-floods in our colonies of the Arctic Regions, our

---

गायन्ति देवाः किल गीतकानि धन्यास्तु ते भारतभूमिभागे ।
स्वर्गापवर्गास्पदमार्गभूते भवंति भूयः पुरुषाः सुरत्वात् ॥ २४ ॥
जानीम नैतत्क वयं विलीने
स्वर्गप्रदे कर्मणि देहबन्धम् ।
प्राप्स्याम धन्याः खलु ते मनुष्याः
ये भारते नेंद्रिय-विप्रहीनाः ॥ २६ ॥
(Vishnu Purána. 2. 3. 22/26).

दक्षिणेन सरस्वत्या दृषद्वत्योत्तरेण च ।
ये वसंति कुरुक्षेत्रे ते वसन्ति त्रिविष्टपे ॥
(Mahá. Bh. III. 81, 204–5).

(1) In fact, Bhárata-Varsha (India), or for the matter of that A'ryávarta, was, as thought by our ancestors, and as has been admitted even now to be, the blessed land of the Sapta-Sindhus, which has been watered by the most renowned Seven Rivers with the Sarasvati the dearest of them all, and which was the scene where Indra, who had achieved his first exploits after conquerring his enemy and killing Vritra, was requested to favour the A'ryas, and punish the non-sacrificers.

(2) It has been the region watered by the most sacred river the Saraswati, which was, with just pride, considered to be *the Scene of creation*, and the region where life had first commenced. (*vide pp.* 99 @ 102).

(3) It has been the Land of the brave and the pious, of heroism and enterprise, of commerce and trade, of

ancestor Manu thought of sailing to the South, owing to the Deluge and the thick sheets of Ice that were fast covering the Arctic Regions, he had in his memory this very Northern Mountain (एतमुत्तरं गिरिम्), meaning thereby the

---

science and art, of virtue and greatness, of countless medicinal herbs and plants.

(4) The Land, encircled by the ocean, and fed by the renowned Seven Rivers, such as the great Indus, the Ganges, and others, and abundantly supplied with grain and food-stuffs of all kinds.

(5) The Land, where the most ancient A'ryans, our *Primitive forefathers* were born and had lived, and where the Gods overthrew the Asuras and evil doers.

(6) The Land, on which extended the Himálayas, the highest Mountains in the world, as also the most beautiful Forests.

(7) The Land of Sacrifices and sacred pleasures, of Valour and Renown, of Patriotism and Self-Sacrifice, of Virtue and Kindness.

(8) The Land of six-seasons, and of equal days and nights. (Vide ante *p.* 74, note 8).

(9) The Land of Choice-Regions, created by God, and considered to be *the scene of creation,* and *the very Heaven on Earth.*

This sort of extreme love, reverence, and fond affection for A'ryávarta—*the Mother-Country and the renowned Land of the Seven Rivers* (सप्तसिन्धवः)—made our ancestors keep constant and uninterrupted communication with it, while they were sojourning in the distant colonies of the Arctic Regions, and elsewhere.

Himâlaya Mountain to the North of A′ryâvarta, where he and his older ancestors were born and had lived first, and seen the *Mountain–wall* to their North, proves that we did not at first belong to the Arctic Regions, nor to Europe, nor to the Central Asiatic Plateaux, but had gone there from A′ryâvarta, for colonization, for conquests abroad, and for quenching the thirst of curiosity.

Moreover, this appears to have been corroborated by the remarkable Deluge–story narrated in the Shata-Patha-Brâhma*n*a, in which the Fish is said to have sailed to the South and landed Manu in a ship on a peak of the highest mountain, known to him as *the Northern Mountain*, because it was to the North of A′ryâvarta, "*the Aryan Cradle.*"

स औघे उत्थिते नावमापेदे । तं स मत्स्य उपन्यापुप्लुवे । तस्य शृंगे नावः पाशं प्रतिमुमोच । तेन एतमुत्तरं गिरिमाधिदुद्राव । (Sh. P. Br. I. 8. 1. 5).

However, apart from this, and to crown the whole, there comes in unexpectedly the testimony from an altogether independent source, and the Avestic evidence affords, beyond all doubt, strong corroboration and supports the theory of the A′ryâvartic Home and the A′ryan Cradle in the Land of the Sapta-Sindhus, or the famous region watered by the World-renowned

Seven Rivers of A'ryâvarta. For, in the first Fârgard of the Vendidad, there appears the description of the sixteen lands created by Ahurmazd, the Supreme God of the Iranians, in which, among others, the regions of (1) Harahwaiti, (2) Hapta-Hendu, and (3) Ranghâ seem to be included, and these respectively appear to have been identified with the (1) Sarasvatî (सरस्वती), (2) the Sapta-Sindhus[1] (सप्तसिन्धवः), and (3) the Rasâ (रसा) rivers, which appear to be distinctly mentioned in the Rig-Veda (X. 75-5/6).

Now, from the mention of these rivers in the Vendidad, the Sacred Scripture of the Iranians, the latter seem to have been intimately acquainted with the rivers, the mountains, the surroundings, in fact the whole face of A'ryâvarta, where, owing to religious schism, two parties having been formed, the weaker (the Iranians) had to leave A'ryâvarta and go to the northern regions, pursued by the stronger party, the Indo-A'ryans, who ruled the country, and as such, were in a position to dictate terms to the former. And the flight of the Iranians from the Mother-

---

1 The Sapta-Sindhus or the seven celebrated rivers of A'ryavarta are the Ganges, the Yamuná or the Jamna, the Saraswati, the Satlaj (Shutudri), the Râvi (Parushni or Iravati), the Chenab (Chandra Bhágá or Asikni), and the Sindhu or the Indus. (The Author).

Country A′ryâvarta to Irân, which they subsequently adopted, and from which they derived their name, could be distinctly traced in the Parsi Scriptures—the Avesta, the Zend, the Vendidad, and the Gâthâs—with the graphic description in lively colours.

For, in the Gâthâ (Ustanvaiti), the leader of the party (Zarathustra) says :— "Two armies have been gathered for a combat, in silence". { 2 (44)-15 }.

In the combat, however, the leader and his party having been defeated, he was compelled to leave A′ryâvarta with his camp and followers. And then, very naturally enough, he had given vent to his innermost feelings, and cried aloud, "To what country shall I go? Where shall I take my refuge? What country is sheltering the master (Zarathustra) and his companions? None of the servants pays reverence to me, nor the wicked rulers of the country" (*i. e.* the Indo-A′ryans of A′ryâvarta, who had expelled the leader of the defeated party from A′ryâvarta, the once common Mother-Country, after the schism). { 4 (46)-1 }.

"I know that I am helpless. Look at me being amongst few men. For, I have few men (as I have lost my followers or they have left me). I implore thee (Ahurmazd the wise) weeping, thou living god ". * * * * } 4.(46)-2 {.

• " The sway is given into the hands of the priests and prophets of idols, who, by their atrocious actions endeavour to destroy the human life." * * * * } 4 (46)-11 {. (*vide* Dr. Martin Haug's Religion of the Parsees. Edition 1862. *pp.* 152, 155, 156, 157, from which these extracts have been taken).

All this evidently refers to the religious schism, the subsequent fight, the persecution, and finally the expulsion from A'ryâvarta of the leader and his followers, who thereafter took shelter in Irân and had even gone as far as the Arctic Regions, which they had colonized along with us (the Indo-A'ryans), in the latter part of the Tertiary Period and before the advent of the Great Ice Age, when the climate of the place was mild and genial. For, the Vendidad expressly says : (40) " Once a year, one sees there (*i. e.* in the Arctic Rgtions) Stars, Moon, and Sun, rising and setting ".

(41) " And they think a day what is a year". (*vide* Vendidad, Second Chapter, and Dr. Haug's Parsee Religion. Edition 1862. *p.* 205).

But, six months' day and six months' night is the sure characteristic of the Arctic

Region. It seems, therefore, that the Indo-A′ryans, as also Iranians and other A′ryan branches of the *parent* Indo-A′ryan stock had, after leaving A′ryâvarta—their cradle—, once remained in the Arctic Colonies for a considerable time, in the Pre-Glacial Period, when, all of a sudden, thick sheets of Ice having spread all over the northern regions, and especially occupied the higher latitudes, the Indo-A′ryans returned to their Mother-Country A′ryâvarta—, keeping before their view the highest Himâlayan Mountain (एतमुत्तरं गिरिम्), from which they had marched in all dircetions, as the chief *Guiding Mark* in the great Ice-floods ; the Iranians retraced their steps to their adopted country Irân ; while the other A′ryan branches settled in countries which were free from the ravages of Ice, or where they could find refuge. (*vide* ante *pp.* 74-77, 80, 81, and next *p.* 82).

Now, it appears from the Avestic evidence, that long before the advent of the Glacial Period or the Pleistocene Epoch, clear signs were already perceived of the approach of hard winter, deadly frost, and heavy floods of Ice, at no distant date ; and even *a prophetic announcement* was in due course made, to the effect that, a destructive frost would ensue, which would cover mountains and the face of the Earth.

For, the Vendidad says, "Ahurmazda spoke unto Yima: "O happy Yima Vivanhana! Upon the world of animated beings, the evils of winter will come, and consequently a strong deadly frost." (Vendidad. *ch* II. *vide* Dr. Haug's Religion of the Parsees. Edition 1862. *p.* 204). And evidently, this hard winter or deadly frost was nothing more than the advent of the Glacial Period at the close of the Tertiary Epoch, of which we have given particulars in the fifth chapter.

Thus, both the R*ig*-Vedic and the Avestic evidence, coupled with the scientific testimony derived from hard geological facts, sufficiently proves that our hoary R*ig*-Vedic ancestors lived during the Tertiary Epoch; that they were autochthonous in A'ryâvarta, had colonized the Arctic regions, and made extensive settlements there, as also elsewhere; that they had witnessed the Great-Ice-Age; that at the advent of this Glacial Epoch, they were compelled to abandon the once genial Arctic colonies, owing to sheets of Ice and snow having capped the mountains and spread over the higher latitudes; that it was then that the Indo-A'ryans, who belonged to, and were born in, A'ryâvarta, and had also kept uninterrupted communication with this *their Mother-Country*, during their sojourn in the genial Arctic Colonies, finally returned Home; that the Indo-Iranians remained in their adopted country Irân; and that other minor A'ryan branches endeavoured to settle in the regions which could afford them shelter.

We shall now return for a while to the several queries mentioned before (ante *pp.*30-31), and try to answer them one by one. The irrefragable evidence cited in the preceding pages, conclusively proves, (1) that the oldest of our Vedas the R*i*g-Veda—which by the bye is also the most ancient document in the world—seems to be as old as the Tertiary Epoch, at any rate a portion of it, since some of the utterances of the R*i*g-Vedic Bards appear to show, beyond all manner of doubt, that our R*i*g-Vedic R*i*shis *had personally seen all the Arctic phenomena—viz.* the long Dawns, the long days, as also the long, dreadful, and tiresome nights, that continued for months together; (2) that our R*i*g-Vedic ancestors were older than the Quaternary Period; (3) that they belonged to the Tertiary Era; (4) that they had seen the Great-Ice-Age; (5) that the fact seems to have been supported and fortified by the R*i*g-Vedic and Avestic evidence; and (6) that, above all, even the testimony of the science of Geology has proved (ante *pp.* 5,6,32,33) the existence of the Tertiary Man.

Before concluding, I may with advantage make some observations and give opinions of eminent geologists and *Savants*, in view of giving to the Reader some idea as to the time that has elapsed since the close of the Tertiary Epoch and that of the Glacial Age. According to Dr. Croll, the Glacial Period must have begun 240,000 years before, and ended, followed by the Quaternary or Post-Glacial period, about 80,000 years ago, having lasted for nearly 160,000 years, with alterations of milder and even tropical temperature. (*vide* Dr. Croll's *Climate and Time*, and *Climate and Cosmology*). Many American geologists, however, are of opinion that the close of the last Glacial Epoch must not have taken place at such on earlier date, and assign a period of only 8,000 years to the close of the last Glacial Epoch. But, Professor Geikie and other reputed geologists think otherwise, as they consider that there were five Glacial and four inter-Glacial Epochs, and that the total duration thereof must have extended to about 80,000 years.

Lyell, who had visited the Niagara-falls in 1841, after studying and carefully reconsidering all the data in the matter of investigations, by which geological time could be measured in years, concluded that the time since the close of the

Glacial Epoch was probably 31,000 years. And, curiously enough, Professor J. W. Spencer arrives at the result, apparently almost identical with that of Sir Charles Lyell, *viz.* 32,000 years. While, as regards the time that has elapsed since the commencement of the Post-*Vindhyan* or Cambrian Epoch, John W. Judd says, " Indeed, the results actually arrived at, by different observers, for the period of time which has elapsed since the commencement of the Cambrian to the present day have varied from 70,000,000 years (Walcott) to 6,000,000,000 years (Mc Gee)." (*Vide* The Student's Lyell. By John W. Judd. *p.* 592. Edition 1896).

CHAPTER III.

Modern Geological Researches of the West.

We have shown in the previous chapters what progress was made in Geological Researches by our great hoary ancestors; and the innumerable references to geological formations in the Puranic and Vedic Literature, clearly indicate that a very lively interest was taken in the subject, and the science was cultivated with unabated zeal and renewed vigour in the Vedic times, notwithstanding the insuperable difficulties the Vedic R*i*shis had to encounter, especially as they had to stand on their own legs altogether unaided.

From the trend of thought and previous discussion, the Reader will have perceived that the Earth was supposed to be at first in a gaseous state or igneous fusion (ante *pp.* 13-17), and that subsequently, part of heat of the globe having been diffused in the surrounding space, refrigeration had gradually proceeded. This naturally caused the aqueous vapour in the atmosphere to condense, and occasioned the fall of rain. The water thus accumulated gave rise to what is called *the first Thermal Ocean*, in which the granite crust having been formed and then partially broken, land and mountains began to rise above waters. Subsequently, after lapse of ages, the thermal deep as also land and mountains began to cool down, and life became visible first in sea-weeds, then in trilobites, fishes, reptiles, quadrupeds, and others, the vital gradation having by process of evolution progressed from simpler conceptions to complex types, till perhaps the last and marvellous production of man,—the crowning piece of creation,—appeared on Earth.

The Reader will have also become aware, that *Creative Energy*, according to the Taittirîya A′ra*n*yaka, was in existence ever since the commencement of time, but was only dormant in the fluid (ante *p*. 18); and that, vitality had commenced first with *the weeds or the lotus*, according to the Taittirîya Bráhma*n*a (*p*. 18). But, according to the Taittirîya Upanishad, life appears to have begun with the moss, *herbs*, and small plants, and then progressed in gradation from small and simple organic bodies to great and complex productions (ante *pp*. 15-16).

And, it may not be out of place here to mention, that a reference to *the Lotus* having appeared in waters (Taitt. A. I. 23. 1; T. Br. I. 1-3-5), and to the *Herbs* on Earth (Taittirîya

Upanishad II. 1) ; or to the *Fish* (V. P. I. 4. 7, foreshadowed in the Shata Patha Bráhmana I. 8. 1. 1 (@ 6), or to the *Tortoise* (T. A. I. 23.4; Sh. P. Br. VII. 5.1.5), or to the *Boar* (Sh. P. Br. XIV. 1. 2. 11), The *Man-Lion* (M. Bh. XII. 339-103,104), the *Dwarf-Wáman* or *Man* (Agni. P. IV. 7), &c., as having appeared on Earth in various forms and made this creation, only indicates that after the Earth had cooled down and life appeared thereon, the various life-types of the kind, *viz.* the lowly weeds represented by the *Lotus* or the *Herbs*, the aquatic beings represented by the *Fish*, the reptiles represented by the *Tortoise*, the mammalia represented by the *Boar*, the intermediate order between beast and man represented by the *Man-Lion*, or according to the Western Geologists and Scientists by the Ape, and finally *the man* represented by the *Dwart-Incarnation* or *Wâman*—the crowning piece of creation—had made their appearance on Earth, in the evolution and progressive gradation of vitality, from time to time, and in geological sequence.

Now, with reference to the type of the *Man-Lion*, alluded to before and supposed by our Vedic as well as Paurânic geologists to be the precursor of Man-kind, and probably as a *connecting link* between *the brute intelligence* and *the reasoning powers of man*, I may here observe in passing, that the Man-Lion was, in all likelihood, supposed to be one of the *lost types* of the anthropoid primates; or, it might be, that the geological knowledge of our ancestors was imperfect. But, in all probability, the *fossils* of such an *anthropoid primate* were found by our older ancestors, as some details appear to have been given in respect of the powers and qualities of the *Man-Lion*. For, there is extant the Upanishad, known as *Nri-Simha-Pûrva-Tápanî*, and the text of the same, be it noted, runs as follows :—

अथ कस्मादुच्यते नृसिंहामिति यस्मात्सर्वेषां भूतानां ना वीर्यतमः श्रेष्ठतमश्च । सिंहो वीर्यतमः श्रेष्ठतमश्च तस्मान्नृसिंह आसीत्परमेश्वरो जगद्धितं वा एतद्रूपमक्षरं भवति (नृसिंहपूर्वतापनी. २ उ० खं० ५).

This may be rendered into English as under :—" Why is Man-Lion so called? Because, man is superior to, and is the most powerful of all creatures. So is lion. Therefore, the Man-Lion has been supposed to be the highest god, a blessing to the world, and an ever-lasting manifestation. " (*Nri. Pûrva-T. Up*. II. 5).

In the absence of evidence, however, the Western geologists would probably ridicule the idea of the intermediate link being *the Man-Lion*, as they have been accustomed to think more of *ape* or *chimpanzee* instead. But, as new fossils and life-types, in regions far and near, explored and unexplored, are being discovered and brought to light with the rapid march of the Science of geology, there need be no disappointment in this respect. For, as said by our great Sanskrit Poet Bhavabhûti in his dramatic Work *Málati-Mádhava*, "Time is endless, and the extent of the Earth is immense" (कालो ह्ययं निरवधिर्विपुला च पृथ्वी). Therefore, with Sir Charles Lyell, "may we hope to meet, hereafter, with the lost types of the anthropoid Primates."

Now, according to the idea of Western geologists, there might perhaps be a mistake in the conception and investigation of our Vedic geologists. But, even taking for granted that *Man-Lion* could not possibly be *the missing-link* or the precursor of Man, in the evolution and development of the life-types, I may here take the liberty respectfully to ask, "Who does not err in the crude state of the science? Or, what theories are not advanced in the light of facts and knowledge available at the time? And how many theories have been *exploded* that once appeared to be *sound and charming*, nay, universally *applauded and accepted*, even by eminent men of Science?"

Be that as it may, it seems, beyond any manner of doubt, that our Vedic Fore-fathers had taken very deep interest in the studies of Geology, and made, as far as lay in their power, careful researches in the science. Nay, they were Evolutionists *par excellence*, and I may say this without fear of contradiction. For, our voluminous Vedic Literature has conclusively proved this, and even European scholars[1] of repute have admitted the fact.

Proceeding further, and going back to the earliest times of which we possess any record, we find very startling discoveries made by the R*i*shis of the R*i*g-Veda period (*vide* ante *p.p.* 13 @ 22). I say *startling*, because the time that has elapsed since then, though not actually incalculable, is simply immense, and the R*i*shis had even then

1 For, Sir Monier Williams says:—"......A system which is in some respects almost identical with that thought out by Spinoza and the profoundest thinkers of modern Europe. Indeed, if you will pardon the anachronism, the Hindus were Spinozites more than two thousand years before the existence of Spinoza, and Darwinians many centuries before Darwin, and Evolutionists many centuries before the doctrines of Evolution had been accepted by the Huxleys of our time, and before any word like Evolution existed in any language of the world." (The Religion of the Hindus. The Indian Magazine and Review. No 249. September 1891).

declared a few Geological facts of great magnitude either by intuitive perception, or by long, arduous, and persevering study, over hundreds of centuries befor Steno[1] or Leibnitz,[2] Lehman[3] or Werner,[4] William Smith,[5] or other lesser satellites of the 17th, 18th, and 19th centuries of the Christian Era, could throw any ray of light on this very important but abstruse and scientific subject (*vide* ante *pp.* 13-24).

From this point of view, therefore, a peep into the Geological past of the West, or a glance at the previous history of the development of Geological science in Europe, will not only be interesting, but will in every way be highly instructive. I shall, accordingly, endeavour to give, in brief, the early progress of geological thought and speculation in Europe.

The first European satellite that had appeared above the horizon, and advanced certain theories in respect of geological rocks, was Steno. He was an Italian, and had, in 1669, introduced

1 An Italian. A. D. 1669.
2 A German. A. D. 1680.
3 A German. A. D. 1756
4 A German. Close of the 18 century A. D.
5 An Englishman. A. D. 1809.

(1) *Primary rocks*, that is rocks void of fossil, and contemporaneous with the creation of the Earth, and

(2) *Secondary rocks*, that is rocks supposed to be fossiliferous, and formed after the creation of the Earth.

Subsequently, the great Leibnitz had, in 1680, divided rocks into

(i) *Stratified*, supposed to be the product of deposition in water, and

(ii) *Unstratified*, believed to be the result of igneous fusion,

and propounded a theory on the igneous origin of our planet, which supposed the Earth to be at first in a state of igneous fusion.

More than half a century later, Lehman, a German miner, proposed in 1756 to divide rocks into three classes: The first and the oldest to be called (i) *Primitive*, comprising hypogene, or plutonic and matamorphic rocks; the next to be termed (ii) *Secondary*, comprehending the aqueous or fossiliferous strata; and (iii) the Remainder or the *Third class* corresponding to the alluvium—ancient and modern—which he referred to local floods and the deluge of Noah.

Now, this primitive class or formation contained no organic remains. Its origin may have been purely chemical, evidently antecedent to the coming into being of vitality, and probably co-eval with the birth of the world itself. The secondary formations contained sand, pebbles, and organic remains as well as mechanical deposits, produced after the planet had become the habitation of plants and living beings.

Werner came next, and he attempted to improve Lehman's classification, by intercalating a class, distinguished by him by the name of "*Transition-formation*," and inserting it between the *Primitive* and *Secondary*. Werners divisions may, therefore, be given as follows :—

(1) *Primitive.*
(2) *Transitional.*
(3) *Secondary.*
(4) *Alluvial.*

I may here state, that the term '*Transition*' was applied to the older palaeozoic strata, as indicating a transition or passage from unfossiliferous to fossiliferous conditions.

The aforesaid theory of Werner's assumed, that the globe had been at first invested by an universal *Chaotic Ocean*, which held the materials of all rocks in solution. This was called the "*Neptunian Theory*," and for many years, it enjoyed much popularity.

In the writings (A. D. 1788-1795) of Hutton of Scotland, a contemporary of Werner, and in those of Play-fair, the germ of metamorphic doctrine is found. Subsequently, and after much controversy, the theory of the igneous origin of trap and granite made its way into general favour, and to reconcile the old doctrine with this new theory, the following hypothesis—*already anticipated by our Vedic ancestors in a great measure*—was substituted for that of the Neptunists :—

(1) The Earth was supposed to be in a state of igneous fusion, until part of the heat having been diffused into the surrounding space, the surface of the fluid consolidated and formed a crust of granite. (*Vide* Statement No. i, Chapter iv of this work).

(2) Subsequently, refrigeration proceeded in time, and this caused the aqueous vapour in the atmosphere to condense.

(3) This condensation caused the fall of rain, which gave rise to the first *Thermal Ocean.*

(4) The temperature of this boiling sea was, in the very nature of things, too high to be favourable to the aquatic beings inhabiting its waters. These, therefore, were devoid of fossils. Moreover, they were highly crystalline, and consequently produced gneiss, mica, schist, and the rest.

(5) The granite crust having been partially broken up, land and mountains began to rise above the waters; while the rains and torrents that ground down rocks, caused the sediment to spread over the bottom of the sea.

(6) The boiling waters, and the burning land and mountains gradually cooled down to a degree, sufficient to sustain life, and then vital gradation and progress commenced from simpler conceptions to more highly organized orders. (*Vide* Statements No. ii, iii, iv, Chapter iv of this work).

About the beginning of the 19th century, that is A. D. 1809, William Smith, the father of English Geology, proceeded more scientifically in the matter, while Sir Charles Lyell's *Principles of Geology* appeared in 1830-33.

Professor John W. Judd also names other satellites who had worked in the field of Geology, and says, " Certain Arabian writers of the tenth century, and afterwards the philosophers of Italy in the sixteenth, seventeenth, and eighteenth centuries–Lionardo da Vinci, Fracastoro, Steno, Scilla, Lozzaro, Moro, Donati, &c—with Hooke, Boyle, and Michell in England, Palissy and Buffon in France, and Raspi, Fuchsel, and others in Germany, laid the true foundation of a geological science, based on the observation of the existing order of nature." (The Student's Lyell. *p*. 5. Edition 1896).

CHAPTER IV.

Geological Researches of the East and the West compared.

In the present chapter, I introduce a few important facts to enable the Reader to get a view of the other side of the shield, and to see for himself how the Vedic Rishis were immeasurably in advance of the times, and had anticipated the Western geologists and scientists, in the deep researches they had made even at that distant date, not to say in the remotest period of time, when all the other ancient nations, including even the Egyptians, the Babylonians, and the Assyrians, were still in the dark.

For, says Thornton in his History of India, "Ere yet the pyramids looked down upon the valley of the Nile,—when Greece and Italy those cradles of modern Civilization, housed only the tenants of the wilderness,—India was the seat of wealth and grandeur."

I had, in the second chapter (*pp.* 35 @ 83) made passing reference to the A'ryan cradle in A'ryâvarta from the geological point of view, and shown therein, that the origin of life was in the region of the sacred river Sarasvatî, situated in A ryâvarta—the Land of the Seven Rivers. Besides, a careful glance at the Vedic works, and especially a perusal of the Rig-Veda, will convince the Reader, that the Sarasvatî was not only deemed by the Rig-Vedic Rishis to be the most sacred river, but was actually supposed to be *specially their own*, as they often used to say with extreme delight, "*our sacred Sarasvatî*" (पावका नः सरस्वती । Rig-Veda I. 3. 10), evidently establishing thereby their exclusive claim on her, even debarring others from participation in her favours, and enjoying themselves the privilege denied to others by Nature herself. It was obviously for this reason, that the river Sarasvatî was *ever associated with the utmost sanctity*, and as such, sacrifices were always performed in the regions and on the banks of the said river: (ऋषयो वै सरस्वत्यां सत्रमासत । Aitareya Br. I. 19; माध्यमाः सरस्वत्यां सत्रमासत । Kaushîtakî Br. XII. 3). While in the Rig-Veda, the said river was even eulogised as *the best of Mothers, the best of Rivers*, and *the best of Goddesses* : (अंबितमे नदीतमे देवितमे सरस्वति । Rig-Veda II. 41. 16).

And all these praises seem to have been lavished on her, evidently for the reason that she was supposed to be the site not only of *the A'ryan Home* and *the Human Cradle*, but was moreover, considered to be *the very region of the Origin* of life, or of vitality itself. Because, the river appears to have been thus addressed :—

त्वे विश्वा सरस्वति श्रितायूंषि देव्याम् ।

(R*i*g-Veda. II. 41.17).

"O Sarasvatî, all life is in thee, who art divine."

Now, this poetic effusion of the Bard does not seem to be an out-pouring of some hackneyed theme, or a common-place thought, or any meaningless expression, but appears to be an original idea in the researches of geology, as the poet had apparently hit upon some geological discovery, that *vitality had first come into play in the region of the river Sarasvatî.* This idea, therefore, being of great moment, and practically of primary importance, I beg to give the translation of the verse, as rendered into English by Oriental and Occidental scholars of note.

"All life is in thee the goddess, O Sarasvati." R*i*g-Veda II. 41. 17. (S. P. Pandit).

"In thee, Sarasvati, divine, all generations have their stay." (R. T. H. Griffith).

"In thee, Sarasvati, who art divine, all existences are collected." (H. H. Wilson).

The import speaks for itself, and to all intents and purposes, as also from the previous context (*vide* R*i*g-Veda II. 41.16., ante *p.* 99), and the Vedic researches in geology (*pp.* 12-24), it conveys the idea that the river Sarasvatî was supposed, from the geological point of view, to be *the scene of vitality* and *the region where life had first commenced after the Earth cooled down*, as the verse says that, 'all life, or generations, or existences, are in the river Sarasvatî.' For, it seems from the keen investigations in geology, continued during the Vedic period (ante *pp.* 12-22), that the R*i*shis had probably come across some fossils in the beds and regions of the river ; and as these were considered to be the earliest life-types, it was naturally thought that *vitality had its origin in that region*, and that, therefore, the very source of life was considered to be in the river itself (त्वे विश्वा सरस्वति श्रितायूंषि देव्याम् । R. V. II. 41.17.).

Let us, however, turn for a while to modern investigations in the matter, and see whether there is any foundation for the Vedic supposition in respect of *the origin of life having been in the region of the river Sarasvatî*, or thereabouts. There is hardly any doubt that the most ancient life-types are found, as observed hereafter (*vide* next *p.* 102), in A'ryâvarta or Northern India, and even Western geologists bear testimony to, and throw their weight of evidence in support of, the fact. For, in this respect, Mr. Medlicott Superintendent Geological Survey of India, says :—" And the most ancient form of life occurs (in India) near the Eastern end of the hills", *viz. the Salt Range of the Punjab.* (Manual of Indian Geology. *p.* XXIV). Moreover, subsequently, and a little later, the same authority states, " still further East too, in the North of Kumaun, Silurian fossils have been discovered in considerable quantities." (Ibid. *p.* XXV).

But more than this, " Noetling has recently described a series of strata as underlying beds containing (the remains of the genus called) *Olenellus*, in North West India ;" and he confirms the conclusions of Waagen that this series of strata contains fossils. He also asserts with confidence that these are of very ancient epoch, and even *of older age* than *the Lowest Cambrian*, that is *Pre-Cambrian*, meaning thereby to say that the fossils belong to *the Vindhyan* Era, (*vide* the Students' Lyell, edited by J. W. Judd, 1896. *p.* 438 ; The Imperial Gazetteer. Indian Empire vol. I. *p.* 55. Ed. 1907).

Thus, the researches of our hoary ancestors that describe A'ryâvarta to be *the scene of* *mitive vitality*, and the region where life had first originated, seem not only within the mark, but marvellously correct in the main, as they have stood the test of ages, and the facts have been corroborated by independent testimony, not to say matter-of-fact foreign evidence. We shall, therefore, turn our attention to other particulars which are of greater moment, and as such, demand our close observation.

In the first chapter, I had occasionally referred to the various rock-formations, which included also the different vital gradations. But, as it is intended in the present chapter to institute comparison between geological researches of the East and the West, I shall here endeavour to show in brief the principal points of resemblance, and introduce four statements, for the elucidation of my allegation and the argument that our Vedic Fore-fathers were acquainted with the science of geology, as they seem to be aware of the Earth's primary gaseous condition, its subsequent liquid state, its final solid form, nay, its Paleontology and successive periods of animate creation in the Palæozoic, Mesozoic, and Cainozoic or Tertiary Epochs.

THE AZOIC EPOCH.

The Earth in *its primary* or *gaseous state*, *its subsequent molten* or *liquid condition*, and *its final solid crust*.

STATEMENT No. I.

The Vedic Texts and Authorities.	English Translation thereof.
१ (अ) प्राचीनान्पर्वतानदृंहत् । (इन्द्रः). ऋ० वे० २.१७.५.	(*a*). He (Indra) made the moving mountains firm, that is solid; (moving, because, the mountains were formed of the mass of liquid matter).
सायणभाष्ये :— प्राचीनान् इतस्ततः प्रकर्षेणांचतोगच्छतःपर्वतान् ।	Wilson translates this as follows :—"He fixed the wandering mountains." (Rig-Veda. II. 17.5. Ed. 1854, vol. II. *p.* 251).
(आ) यः पृथिवीं व्यथमानामदृंहद्यः पर्वतान्प्रकुपितानरम्णात् । (ऋ० वे० २-१२-२).	(*b*) Who (Indra) made the trembling (because moving to and fro, like water, on account of the liquid state of the) Earth firm, and made the mountains solid (that were rolling at first, when in a molten condition). Rig-Veda II. 12,2.*

* "He, who fixed firm the moving Earth; who tranquillised the incensed mountains." (Wilson's translation. Ed. 1854, vol. II. *p.* 236).

The verse, however, refers beyond doubt, to the molten state of the Earth, and its subsequent solid condition and hard crust.

The Vedic Texts and Authorities.	English Translation thereof.
(इ) येन...पृथिवी च दृढा । (ऋ. वे. १०. १२१. ५).	(*c*) "By him, ... the Earth is steadfast." (Griffith). (Rig-Veda. X. 121. 5).
(ई) अपामुपस्थे विभृतो यदा-वसत्......। (अग्निः) । (ऋ. वे. १. १४४. २).	(*d*) The fire was then on the lap of waters, and dwelt extended therein. (This means that, before the Earth became firm or solid, it was in a calorific molten state. That is to say, it was all a *thermal deep*. and this expression seems to have been couched in the aforesaid words of the text). (R. V. I. 144-2).
२ (अ) आकाशाद्वायुः । वायो-रग्निः । (तैत्तिरीयोपनिषद् २-१)	2 (*a*) From the sky, that is, from ether, there was wind which produced fire. (This means that, in the infinite space, there was wind; and this produced fire.) (Taitt. Up. II. 1).
(आ) अग्नेरापः । (तै. उ. २-१).	(*b*) The fire produced water (in course of time). (Taitt. Up. II. 1).
(इ) अद्भ्यः पृथिवी । (तै. उ. २-१).	(*c*) From water came out lands (gradually). (Taitt. Up. II. 1).
३ (अ) इदं वा अग्रेनैव किंच नाऽऽसीत । न द्यौर सीन्न पृथिवी । नान्तरिक्षम् । तदसदेव सन् मनो कुरुत-स्यामिति । तदप्यत । (तै. ब्रा. २. २. ९. १).	3 (*a*) " At first this (that is, the universe) was not any thing. There was neither sky, nor earth, nor air. Being non-existent, it resolved, 'let me be.' It (then) became fervent."

The Vedic Texts and Authorities.	English Translation thereof.
	(Taitt. Br. II. 2. 9. 1). *N. B.* I may here gratefully note, that the translation of the original text has been borrowed from Muir's Original Sanskrit Texts. Vol I. P. 28. Second Edition. (*The Author*).
(आ) तस्मात्तपनाद्धूमोऽजायत।...अग्निरजायत।...ज्योतिरजायत।...तदर्चिरिव समहन्यत। (तै. ब्रा. २. २. ९. २).	(*b*) "From that fervour, smoke was produced, ...(as also). fire, ...light, ...and blaze, ... It then became condensed like a cloud." (Taitt. Br. II. 2. 9. 2).
(इ) तद्वस्तिमभिनत्। (तै. ब्रा. २. २. ९. २.). स समुद्रो ऽभवत्। (तै. ब्रा. २. २ ९. ३).	(*c*) "It (afterwards) clove its bladder, and that became the sea" (that is, the *Thermal Ocean*). (Taitt. Br. II. 2. 9. 2, 3).
(ई) यदप्स्ववापद्यत सा पृथिव्यभवत्. (तै. ब्रा. २. २. ९. ४).	(*d*) "That which fell into the waters became the Earth". (*viz.* Prajápati's crystal like tears). (Taitt. Br. II. 2. 9. 4).
(उ) तां (पृथिवीं प्रजापतिः) शर्कराभिरदृंहत् (तै. ब्रा. १. १. ३. ५).	(*e*) And Prajápati "strengthened her (the Earth) with gravel." (Taitt. Br. I. 1. 3. 5). *vide.* Muir's O. S. T. I. 28, 29, 53. 2nd Ed.

Now, it would not be out of place here to remind the Reader, that all ideas of our *Primitive ancestors* and Ancient fore-fathers, whether *geological* or *astronomical*, *philosophical* or *governmental*, *literary* or *scientific*, were ever clothed in the attire of Religion, as this had always to them the all-absorbing interest. And, as observed by Max-Muller, in his "*India. What can it teach us?*" (Ed. 1883. *pp.* 107-108), "it embraced not only worship and prayer, but what we call philosophy, morality, law, and government,—all was pervaded by religion. Their whole life was to them a religion—every thing else was, as it were, a mere concession made to the ephemeral requirements of this life." Naturally, therefore, there was among Vedic Geologists and scientists difference of opinion, which, however, exists amongst scientists of all ages, and which, without detracting its value, only adds to its flavour and importance, as it indicates that the Vedic geologist, philosopher, or scientist, had his own stand-point in the investigations in which he was engaged, and shows, moreover, that his researches were not only independant, but were continued without the slavish spirit of servile imitation.

All the description, therefore, in statement No. I, though clothed in mystical words, relates to the primary state of our globe, its gaseous, liquid or molten condition, and to the Azoic Period, when, owing to the heat of the planet, life could not exist. But refrigeration was gradually proceeding in the very nature of things, and the consequent introduction of vitality after considerable time, had become compatible with its existence. For, the origin of life necesarily implied the fitness of the Earth for its sustenance, and when once the vitality commenced, there was progress from simpler conceptions to more highly organised orders,—that is to say, from the weeds, herbs, and trilobites of the *Palæozoic Era* to the bone-clad fishes of the *coal-measures*, the repuiles of the *oolite*, and the huge mammalia of the *Tertiary Epoch*, with Man as the crowning piece of creation of the Period, as will be presently shown, in brief. We shall, therefore, turn our attention for a while to the palæontology of our Vedic Fore-Fathers, and give a few details of the *Proterozoic* and *Palæozoic* Period, *Secondary* or *Mesozoic* Epoch, and *Tertiary* or *Cainozoic* Era, (following Sir Charles Lyell, the great authority, in the classification of Geological Periods), in the undermentioned statements:—

It may be observed that, one thing at any rate appears to be singularly prominent in the Geological Researches of our R*i*g-Vedic fore-fathers; and we cannot let this fact pass unnoticed, especially as the Vedic researches in

Geology seem to be most convincing in the main fact, that the R*i*shis of the R*i*g-Vedic period had discovered, even at that distant date, that vegetable growth, or vitality in herbs and plants had appeared, or for the matter of that the very growth of life on Earth, first commenced, in the Palæozoic Period, that is, some three Epochs before the advent of Man—the crowning piece of Mammalian order, nay, even of creation itself. Because, there appears an express mention of the fact in the R*i*g-Veda (X. 97. 1), which we shall give *in extenso*, in the next statement No. II. *p*. 112. In short, according to the R*i*g-Vedic and Purânic Geologists, the whole thing in a nut-shell seems to be that, after our Planet cooled down, and acquired a condition compatible with the existence of life, herbs (ओषधीः), and fish (मत्स्यः), including trilobites, appeared in the first Geological Period (पूर्वा जाता), *viz*. the Palæozoic Era; while, in the third or the Tertiary Epoch, the Mammalian order, with the Man or the Gods as the crowning piece of creation, made its appearance (देवेभ्यस्त्रियुगं पुरा । R. V. X. 97. 1). We shall, therefore, for the sake of comparison, and for bringing conviction home to the reader, turn our attention for a while to the Western and modern geologists of note, and see what they have to state in the matter, as our Vedic

Geological ideas and conceptions throw considerable light on the state of Geological Science, during the hoary Ṛig-Vedic Period.

Sir Charles Lyell, the great authority in Geology, makes mention of the *Devonian Flora, viz.* of the remains "of marine plants or fucoids" (*vide pp.* 349, 344. Manual of Elementary Geology. Fourth Revised Edition. 1852). Dr. David Page refers to the *Silurian Flora* and says that, " The Silurian Flora,...consists chiefly of algæ or marine plants. " (Advanced Text-Book of Geology. Edition 1861. *p.* 160). In the same manner, Dr. James D. Dana, the great American Geologist, and Silliman Professor of Geology and Natural History, in Tale College, refers *sea-weeds* and *plants* to the Silurian Epoch (*p.* 395), and fossil plants to the Coal-Era (*p.* 332). He even goes further by observing that, " The earliest land-plants were the highest of the Cryptogams...The vegetable kingdom began with the lowest of its tribes, the Algæ, or sea-weeds, and probably with the lowest of sea-weeds, as far back as the Azoic Age." *P.* 396. (Manual of Geology, 1863. By James D. Dana, L. L. D.). Professor Judd also recognises Ordovician (Siluro-Cambrian) Flora, and says that, " Among these (examples of true algæ) must probably be classed the curious

Calcareous algæ (*Girvanella, &c.*)"... (The Student's Lyell. Edited by John W. Judd, C. B L. L. D. F. R. S. 1896. *p.* 411). While, from the observations of Dr. Noetling, it seems that even pre-Cambrian life, and fossils of older age than the Lowest Cambrian or Vindhyan (विंध्यकालीन), are found in the North-West-India. (Student's Lyell, by Judd. *p.* 438. Ed. 1896.)

I may here observe in passing, while noticing and following our ancient geological traces, that the Vedic and Purânic R*i*shis had probably classified the Indian Geological Epochs in their own way, by finding out and connoting roughly, the prominent life-types of each Geological Period, and naming the Geological Eras by these names. For instance, it was discovered by our Vedic R*i*shis, while making geological Researches, that the Palæozoic Era had commenced with vitality, in the order of appearing *first, in herbs or plants* (या ओषधीः पूर्वाजाता R. V. X. 97.1 ; पृथिव्या ओषधयः। तै० उ० २-१) and then in the *fish*, all be it microscopic, including trilobites &c. This Geological Era, therefore, was said to be the ओषधि-मत्स्य-काल, better known as the First-Incarnation (मत्स्यावतार) in the Purânic times, as religion first entered into everything Hindu, not to say even in their Science (*vide* ante *p.* 14,107). The *vitality* in *herbs* and *fish*, which connoted the Palæozoic Era, was followed by the *Crawling* or *Reptile Life*—the Mesozoic Epoch-which having been best represented, according to the notions of our fore-fathers by कूर्म or *Tortoise*, it was called the कूर्मावतार or *the Tortoise-Incarnation*. This "*Age of Reptiles*" or the Mesozoic Era was succeeded by *the Age of Mammals*, or the Tertiary (otherwise called *Cainozoic*) Epoch; and the prominent life-types of this period, according to the Vedic and Purânic theory, having been supposed to be the mammals of the following kind, *viz.* वराह (*the Boar*), नृसिंह (*the Man-Lion* or the link between *Man* and *Beast*, as already noted (*pp.* 88 @ 90), and वामन (*the Dwarf-Man*), the Epochs were called after these names, and the life-types representing the Epochs were considered to be incarnations (*vide* Details in Statements Nos. II, III, IV, and subsequent explanations).

The Proterozoic and Palæozoic Period.

STATEMENT No. II.

The Vedic Texts and Authorities.	English Translation thereof.
१ (अ) या ओषधीः पूर्वा जाता देवेभ्यस्त्रियुगं पुरा। (ऋ. वे. १०-९७-१).	1 (*a*) "Herbs that sprang up in time of old, three ages (that is three Epochs) earlier than the Gods."[1] (Griffith).
(आ) पृथिव्या ओषधयः। (तै. उ. २. १.)	(*b*) Herbs that appeared on Earth, (after it cooled down). (Taitt. Upanishad. II. 1).

1 I may here remark that Gods are but *men endowed with superior qualities*. And as a writer, Mr. Charles Calloway, M. A. D.Sc., has also observed, while speaking about "*The Unity of the Gods*", that "the Gods were but portions of our human nature."

In this statement, the special mention of the fact, that herbs or weeds had sprung up first (या ओषधीः पूर्वा जाता), some three Epochs before the advent of Gods—the superior human beings—(देवेभ्यस्त्रियुगं पुरा), is of great moment, as it bespeaks but *original investigation and great Geological Research* on the part of our Vedic Fore-Farthers of the hoary past, and compares very favourably even with the accepted modern geological ideas, which also place the Proterozoic flora and fauna in the Palæozoic Era, and the Man in the Tertiary or Coinozoic Epoch. While, in respect of the Fish, I may venture to state from the Mahâ-Bhârata and the Purânic works, that our Geologists of the Epic and Purânic times had found, or at any rate were aware of, the oldest known fossil-fish, inasmuch as the Fish seems to have represented the earliest organic body in their works. This organic body was perhaps the smallest, not to say microscopic, and belonged to the pre-Cambrian period. While, the fish-type, moreover, appears to have been foreshadowed even in the Shatapatha Brâhma*n*a (*vide* I. 8. 1. 1). I may, therefore, quote here with advantage the very important and relevent observations of Professor John W. Judd, in regard to the earliest Indian life-types, as they have been declared by Dr. Noetling to be of the *pre-Cambrian* that is *Vindhyan* Epoch, or of older age than the Lowest Cambrian.

For, "Noetling," says Professor Judd, "has recenly described a series of strata as underlying beds containing *Olenellus*, in North-West India. He confirms the conclusions of Waagen that this series of strata containing fossils, *** and various remains of Annelida, is really of older age than the Lowest Cambrian with Olenellus." (The Student's Lyell, edited by Prof. John W. Judd. 1896. *p.* 438).

The Secondary or Mesozoic Epoch.

STATEMENT No. III.

The Vedic and Puránic Texts.	English Translation thereof.
१ (अ) स यत्कूर्मो नाम । एतद्वै रूपं कृत्वा प्रजापतिः प्रजा असृजत् । (श. प. ब्रा. ७. ५. १.५).	1 (*a*) "As to its being called *Kurma* (a tortoise); Prajápati having taken this form, created offspring". (Sh. P.Br. VII. 5. 1. 5. *Vide* Muir's Original S. T. Vol. I. P. 54. Second Edition).
(आ) कूर्मादिरूपधृक् । (अग्निपुराणम् १-१४). (इ) कूर्मादिकं वपुरास्थितः । (विष्णुपुराणम् १-४. ७).	(*b*, *c*) Assumed the form of a tortoise. (Agni Purá*n*a I. 14; Vish*n*u Purá*n*a I. 4. 7).

Now, in regard to this statement, I may here note, that our Indian Geologists appear to have found that vitality in herbs and fish was followed by amphibious life-types (such as कूर्म or *tortoise*), and that the progress was naturally in the coming into being of such creatures as tortoise, frog, crocodile, &c., which lived in water as also on land. Besides, even modern geologists declare that "amphibians were very abundant in the Trias," which is but a sub-division of the secondary or Mesozoic Epoch. (The Student's Lyell. Edited by Professor John W. Judd. 1896. *pp*. 315, 331).

The amphibians and the crawling or reptile life-types were followed by the mammalian order, and we notice abrupt mention of the same, in the Vedic and Purâ-nic works, as will be perceived from the next statement; although, be it mentioned to the great credit of our Vedic Geologists, that in the Taittirîya Upanishad (II. 1), we discover to a certain extent a connected link in the evolu-tion of Geological rocks, as also of changes in the form of Earth, from the Azoic to the Older Palæozoic or Proterozoic Epoch, and find a typically brief and yet scientific description of the Primary Geological changes and of the state of rocks, to which, by the bye, we have referred to in detail, before. (*pp*. 15 @ 24, 104 @ 113).

The Tertiary or Cainozoic Era.

STATEMENT No. IV.

The Vedic and Puránic Texts.	English Translation thereof.
१ (अ) स (प्रजापतिः) वराहो-रूपं कृत्वोपन्यमज्जत्। (तै. ब्रा. १. १. ३. ६).	1 (*a*) "He (Prajápati) having assumed the form of a boar plunged beneath" (into waters). (Taitt. Br. I. 1. 3. 6). Muir's O. S. T. Vol. I. P. 53, 2 Ed.
(आ) वाराहं वपुरास्थितः। (विष्णुपुराणम् १. ४. ७).	(*b*) Had taken the form of a boar. (Vish*n*u Purá*n*a I. 4. 7).
(इ) वराहो नरसिंहश्च वामनो राम एव च। (महाभारतम् १२. ३३९. १०३-४).	(*c*) The Boar, the Man-Lion, the Man or the Dwarf Wáman, and Ráma (were all the incarnations of God).
(ई) तस्मान्नृसिंह आसीत्परमेश्वरः ......। (नृ. पु ता. उ. २. ९).	(*h*) *Vide p.* 39 above.

All these expressions evidently seem to be an indication of the coming into being of the mammalian order, represented by वराह or the Boar. And although, the idea of our Vedic Fathers, in respect of the geological evolution, progressive creation, and successive grades of vitality, looks rather hazy, and does not appear to have been duly expressed in so many distinct words, still reading between the lines, and connecting the disjointed links of stray thoughts found in the Vedic Literature, it is by no means difficult to give a connected whole; nor is it a hopeless task to get a proper view of the Vedic studies in Geology.

For, from the very brief summary and extremely synoptical view, obtained from the rather confused, disorderly, and stray utterances of our Vedic and Purânic ancestors, it seems obvious that during successive periods of our Planet's history, different life-types, as also groups of plants and animals, appeared on the Earth. That the dominant types of each succeeding Epoch belonged to higher organisation and still more complex structure. That these attained wonderful development. And that they also characterised the area by their numbers and variety of form. For instance, *the Palæozoic* was the period of *flora* and *fauna*, *e. g.* herbs, fishes, &c. (*vide pp.* 109 @ 113). While *Mesozoic* was "*the Age of Reptiles.*" And *Cainozoic* or *Tertiary* was "*the Age of Mammals*". (*Vide* the Student's Lyell. Edited by Professor John W. Judd. 1896. *p.* 447; and above *pp.* 115, 116).

The aforesaid statements, therefore, I think will bring into prominent relief the evolution and the transformation of rocks from their very commencement to the Quaternary period, as found and conceived by the Vedic R*i*shis and our hoary fore-fathers, while the quotations indicated therein will serve to exhibit the various authorities available, in each stage of evolution, from which the texts have been gleaned. I shall, therefore, endeavour to analyse the Vedic Researches of the hoary past, and compare them with those of modern times, with the express purpose of seeing whether our *unaided* forefathers of yore had well hit the target, or were altogether wide of the mark, in the investigation of the science of Geology. And as it is necessary to begin with the origin of our Planet, and then gradually proceed to find out the primary and subsequent formations of rocks, it would, by all means, be more convenient to scrutinise the foregoing statements, and attentively examine them, one by one.

Now, taking the first statement in all its details, we shall at once perceive that it relates to the Primary state of Earth, or at any rate to the condition in which the Globe was supposed to be, at first. What then do the Vedic texts tell us ? Do they give us any valuable information ? Yes. They in rich hue of Golden Letters not only write the history of our Planet and tell its genuine tale, but open up an invaluable mine of wealth that yields rich information in respect of the elements of Vedic Geology, that had remained hidden and unexplored for ages. It is, therefore, obviously necessary to exploit and lay bare such seams as appear most precious.

Our Vedic Forefathers, it seems, first conceived that the planet Earth, with its mountains and solid crust was primarily in an intensely heated gaseous state[1] and igneous fusion; that thereafter, it assumed liquid condition[2] or was in a molten state; and that subsequently, after lapse of ages, it acquired the present solid[3] state, when it cooled down.

---

1 (*a*) तस्मात्तपनाद्धूमोऽजायत । तस्मत्तपनादग्निरजायत । From the fervour (of Prajápati) was produced smoke and gaseous matter, which in turn produced intense heat. (Tai. Br. II. 2. 9. 1. *vide* ante *p.* 106. Statement No. I.)

(*b*) आकाशाद्वायुः । वायोरग्निः । From the sky (Ether) was produced the wind which created heat and fire. (Tai. Up. II. 1.)

2 (*a*) अग्नेरापः । From fire or intense heat was produced water, *viz.* the Thermal ocean. (Tai. Up. II. 1.).

(*b*) आपो वा इदमग्रे सलिलमासीत् । (Taittirîya Br. I. 1. 3. 5).

Muir translates it thus :—" This (universe) was formerly water. fluid ". (*Vide* Muir's Original Sanskrit Texts. vol I. *p.* 53. Second Edition).

This obviously indicates the molten state of our planet; and the fluid was evidently the *Thermal* deep, after the globe that was first in a gaseous condition was reduced in time to the molten state.

3 (*a*) यः पृथिवीं व्यथमानामदृंहत्... । (R. V. II. 12. 2; *vide* ante *p.* 104).

[*See Page 120.*

All this, in other words, evidently means, that the Earth was at first in a state of igneous fusion, until part of the heat having been diffused into surrounding space, it caused the aqueous vapour in the atmosphere to condense. This caused the fall of rain[1] and created *the first Thermal Ocean* (स समुद्रोऽभवत् । तै० ब्रा० २. २. ९. ३) .... Thus, our globe had acquired the liquid condition or molten state. But, the very high temperature of *the Thermal Ocean* or the boiling sea, made the waters extremely crystalline, and from these, granite, gneiss, mica, schist,

---

(*b*) स प्राचीनान् पर्वतानदृंहदोजसा ... । (R. V. II. 17. 5 ; vide ante *p.* 104).

(*c*) येन .... पृथिवी च दृढा ...., (R*i*g. V. X. 121. 5) He (the Lord) made the Earth solid, (that was in a gaseous state and molten condition). *Vide* ante *p.* 105.

(*d*) तां (पृथिवीं) शर्कराभिरदृंहत् । (Taitt. Br. I. 1. 3. 7). He (Prajápati) then made the Earth solid by converting *the Thermal Ocean* into hard gravel (शर्करा). *Vide* ante *p.* 106. (*d, e*).

(*e*) अद्भ्यः पृथिवी । (T. Up. II. 1.)

From the waters (or the liquid state), came out the Earth ; that is, the waters gradually acquired solid condition. *Vide* ante *p.* 106. (*d, e*).

1 स (इन्द्रः) ओजसाधराचीनमकृणोदपामपः (R. V. II. 17. 5.) "He (Indra) with his might ordained the down-ward course of waters". (सायणभाष्ये अपां = मेघस्थितानामुदकानाम्).

&c., having been precipitated, the Earth became gradually solid, while the granite crust having been partially broken up, land and mountains began to rise above waters.

The investigations into the Palæontology and subsequent formations of rocks, also yield marvellous results, as these declare that vitality had not yet come into play, during the Earth's gaseous or liquid state, when the materials of all rocks having been held in solution. every thing was converted into one *Thermal Deep*. Nor could there be life in the solid condition while the Earth's crust was yet too hot for the sustenance of it. But, before we proceed to give further palæontological details of great interest, we must pause for a while and turn our attention to the geologists of modern times, in view of seeing, whether the geological ideas of our Vedic ancestors favourably compare with those of the present age, especially as this commands all the resources of the nineteenth and the twentieth centuries, and receives every sort of available aid that intellect, zeal, and vast progress in the science, could possibly supply; but which, on the contrary, could hardly be expected by our Vedic R*i*shis, during the times of the hoary past and the Tertiary Period.

From the foregoing scrutiny of facts, and the details given in statement No. I (*pp.* 104 @ 108), the reader will have easily perceived, that the central idea of our Vedic Fore-fathers, in regard to the present configuration of the globe was, that

(*a*) it was first in a gaseous state of igneous fusion ; that

(*b*) then, it acquired liquid condition, in which the materials of all rocks were held in solution ; and that

(*c*) subsequently, part of it having become converted into a solid crust as it cooled down, it appeared in the form of lands and mountains.

Let us, therefore, compare this very ancient Vedic idea with the most modern views of the recent West. And I may here, at the very outset, be allowed to offer an humble suggestion that, *in time* as also *in space*, the confines of the Universe, or His works of creation, absolutely lie beyond the reach of mortal ken. Naturally, therefore, many Vedic Scientists and subsequent Hindu philosophers, as also Western *Savants* and geologists, touch at times a most sonorous chord and strike a soft melodious note, by declaring their deep conviction, that so far as the primæval state of our Globe was concerned, *there never was a beginning* to the present order of things. Thus, the great saint, poet, and philosopher,-Jnáneshwara (ज्ञानेश्वर) cf Mahârâshtra—of the 13th Century, says:— " This creation has been in existence from *Eternity* (अनादि). That is to say, *it is without beginning and without end* ". The original verse in Marâthi is as follows :—

तरी सांग पा अर्जुना ।
तुजपासूनि स्थिति या त्रिभुवना ।
हे अनादि विश्वरचना ।
तें लटिकें कायी ॥
(ज्ञानेश्वरी. २-९५).

अनादि सिद्ध हें आघवें ।
होत जात स्वभावें । . . . . ॥
(ज्ञानेश्वरी. २-१००).

(Jnâneshwarî or Bhâvártha-
Dîpikâ. II. 95, 100).

In the same way, the learned Swâmi Vidyâranya,—better known as Mádhava Sâyana, the great Vedic Exegetist and the celebrated (or as some say the brother of the) Prime Minister of king Bukka, of the 14th Century, who held sway over the extensive Empire of Vijayânagar, —writes in respect of the creation of this world, as " being *without beginning and without end* ".

अनादाविव संसारे.... ।

(पंचदशी. १-५९).

(Pancha Dashi. I. 59).

Besides, some R*i*shis (sages) of the Upanishad period declare that, " *That which is*, (that is the Universe), *has been so*, from the beginning of time ".

सदेव सौम्येदमग्र आसीत् ।

(छां० उ० ६-२-१).

(Chhándogyopanishad. VI. 2. 1).

Now, let us for a while turn our attention to the West, and see what the Occidental geologists have to say in respect of the matter. Dr. David Page of the Geological Society asserts that, "in all times past, the same kind of agencies operated on, and modified the rock-materials of the globe." " That then, as now, the world had its oceans and continents, its seas and islands, its lakes and rivers, ... the plains and hills " ...... (Advanced Text-Book of Geology. By Dr. David Page. Edition 1856. *p.* 275). Hutton, in like manner, writes that, " We find no vestige of a beginning, no prospect of an end."

Sir Charles Lyell also contends that, " there are certainly no geological proofs that the granite which constitutes the foundation of so much of the Earth's crust was ever at once in a state of universal fusion." (Elements of Geology. Sixth Ed. *p.* 90).

However, the thirst for knowledge makes us dive deep, and peep into the mysteries of the unknown past; and as the arguments from analogy, in favour and support of a beginning, remain, in the opinion of some Geologists, unassailable and unshaken, the researches become attractive in the extreme. Geologists, therefore, while engaged in these interesting pursuits and the fascinating love of labour, have been compelled, by the evidence of solid facts, derived from the study of the Earth's crust and its rock-formations, to entertain the hypothesis that the globe has "gradually cooled down from a state of molten incandescence to its present temperature" (*vide* David Page's Geology. *p.* 277. Ed. 1856), admitting, however, the fact that science is silent, not to say defective, in respect of direct evidence, though circumstantial evidence goes far enough to sustain the belief of gradual refrigeration and consequent introduction of life at the stage compatible with its existence.

In the same way, the great Leibnitz argued that the whole planet (the Earth) was once in a state of liquifaction by heat or igneous fusion. While at present, four different views have been entertained in regard to the condition of the Earth, of which, therefore, I venture to give here a summary as follows, for the sake of comparing these views with those of the Vedic Fathers of Geology :—That

(1) Our planet is in a molten state, and surrounded by solid crust.

(2) It is practically solid throughout.

(3) It has a thin viscous intermediate stratum, reposing on a solid core, and covered by a solid crust.

(4) It is a globe of gas, enveloped first by an inner molten layer, and next by an outer solid crust. (*Vide* Lapworth's Text–Book of Geology. *pp.* 49, 50. Ed. 1899).

In all these views, however, the hypothesis of the original molten condition or gaseous state of the Planet, and its subsequent slowly cooling process is practically involved. Therefore, this modern and admittedly scientific view of the Western Geologists, belonging to the last, that is, the nineteenth century, very favourably compares even with the most ancient geological ideas entertained by our Fore-fathers of the Upanishad period, about over 4,000 years before (*vide* ante *pp.* 26 @ 29), and even by R*i*shis of the R*i*g-Vedic times, who lived in the Tertiary Period, and as such in the still more remote and hoary past, (*vide* ante *pp.* 82 @ 85). For, they maintained, that the planet Earth was *at first in a gaseous* state *and igneous fusion* (तस्मात्तपनाद्धूमोऽजायत। तस्मात्तपनादग्निरजायत। तै० ब्रा० २–२–९); that it then assumed *a molten state*; the materials of all rocks, *viz.* mountains and earth having been held in solution, and Indra having reduced them to the condition of fluid by his prowess (ओजसा......ऽधराचीनमकरोद्पामपः। Rig-Veda. II. 17. 5; पृथिवीं व्यथमानां......पर्वतान् प्रकुपितान्......R. V. II. 12. 2); that after lapse of ages, it cooled down; and that subsequently, the same became *earth* (अद्भ्यः पृथिवी। तै० उ० २–१), Indra having *converted it into a solid crust* (यः पृथिवीं व्यथमानामदृंहद्यः पर्वतान्प्रकुपितानरम्णात्। R. V. II. 12. 2). *Vide* ante *pp.* 13 @ 18, 20, 21.

Moreover, there appears a distinct statement in the Rig-Veda and the Upanishads, of there having been a *thermal deep* (स समुद्रोऽभवत्। तै० ब्रा० २. २–९–३), or of the fire (अग्निः) having dwelt on the lap of the waters (अपामुपस्थे विभृतः ...... । R. V. I. 144. 2); and this, beyond doubt, indicates a state of things, in which our planet (the Earth) was in igneous fusion and surrounded by watery vapours; or rather, it was in a *liquid* or *molten condition* (अपामुपस्थे विभृतः...... ।). *Vide* ante *pp.* 105 (*d*), 106 (*c*).

Having said so much as regards the Azoic Period, we shall turn our attention to the Palæozoic Epoch, where the comparison becomes still more tempting, and yields results of great moment. The Azoic Epoch, as aforesaid, was a period void of life, as parts of the liquid and solid Earth were yet too hot to sustain vitality. But, as these gradually cooled down, the chaotic condition began to vanish, and life, after lapse of ages, made its appearance therein, in the form of *herbs* and *weeds* (पृथिव्या ओषधयः । Taitt. Up. II. 1), some three Epochs before the advent of Man, the crowning piece of creation (या ओषधीः पूर्वा जाता देवेभ्यस्त्रियुगं पुरा । ......R. V. X. 97. 1. *vide* ante *pp*. 112, 113).

At the beginning, however, the creative energy appears to have had only simple conceptions ; and gradually, it seems to have progressed from simple organic structures to highly organised bodies and more complex orders ; since, the study of Palæontology favours the doctrine of development, or at least of the successive appearance on the earth of beings more and more highly-organised. For, we have first the lowly weeds (पृथिव्या ओषधयः Taitt. Up. II. 1), and then the microscopic fish or trilobite, prior to the mammalian life-types, say of the *boar-order* according to the

Vedic and Purânic geologists (स वराहरूपं कृत्वा-उपन्यमज्जत् । T. B. I. 1.3. 6 ; मत्स्यकूर्मादिकं तद्वद्वाराहं वपुरास्थितः । V. P. I. 4).

And here, by the bye, I may be allowed to observe, that the oldest known fossil-fish, older than the Bone-bed of the upper Ludlow formations, was found in 1859 by J. E. Lee, F. G. S., at Church Hill, in Shropshire, with fossil-shells of the Lower Ludlow or Upper-Silurian group. Moreover, in the Laurentian rocks also, was found *an organic body* in 1859, by Sir W. Logan. This was examined by Dr. Dawson of Montreal in 1864 by means of microscope, and he detected in it a distinct structure of Rhizopod, which was supposed to be the oldest of the known organic remains, and therefore said to be of great antiquity. (*Vide* Lyell's Elements of Geology *p*. 579. Sixth Edition ; David Page's Text-Book of Geology, *p*. 279. Ed. 1856). If however, as remarked by Lapworth, the organic structure called Eozoon or the Dawn animal of Canada is discredited by most geologists, and the preponderance of scientific opinion has long been in favour of regarding it as *a peculiar mineral structure*, imitative of the organic (*vide* Lapworth's Text-Book of Geology, *p*. 182, Edition 1899), then the Pre-Cambrian fossils of North-West India, observed by Dr. Noetling, may be noted with advantage. (*Vide* ante *pp.* 102, 103).

The Vedic idea, therefore, that vitality first began to appear in weeds or herbs (पृथिव्या ओषधयः । Taitt. Up. II. 1), some three epochs before the advent of Man (या ओषधीः पूर्वा जाता देवेभ्यस्त्रियुगं पुरा । R. V. X. 97.1); that gradually the creative energy manifested itself in the form of a fish, all be it microscopic ; that then, it appeared in the form of a *tortoise* ; that subsequently, it made its appearance in the mammalian order of the boar kind ; and that finally, the creative power manifested itself in the form of *Man*, the crowning piecc of progressive creation, after the intermediate or the connecting link of the *Man-Lion*,* seems to be marvellously correct in the main, even from the stand-point of the advanced Western Geological researches of modern times.

And here, by the bye, I cannot resist the temptation to state, that our Indian theory of Incarnations, *viz.* first that of the Fish (मत्स्यः), the second of the Tortoise (कूर्मः), the third of

* For which, the Western Geologists substitute chimpanzee or the ape. *Vide*, however, ante *pp.* 89, 90, 91, where I have offered reasons for the Vedic and Puránic link of the *Man-Lion* (नृसिंह), from the stand-point of our Indian Geologists.

the Boar (वराहः), the fourth of the Man-Licn (नरसिंहः), the fifth of the Dwarf-Wâmana (वामनः) or Man, and so on, is nothing more than a metaphorical expression of the successive phases of geological evolution and development in the ascending order; from the Fish to the perfect type of Man, the crowning piece of creation. (*Vide* ante *pp.* 89 @ 91, 112, 113, 116, 130).

CHAPTER V.

An Episode of the Glacial Period,
&
The Subsequent Quaternary Era.

The preceding chapters will bring into prominent relief, and give the reader some idea of, the fact that our Vedic fore-fathers had knowledge of, and were acquainted with, the elements of Geology and the principal rock-formations. Of course, it has to be admitted, and the fact cannot be gainsaid, that no systematic works nor any treatises appear to have been written on the subject. At any rate, not even one is found at the present day. Yet, there appear side-lights and the scattered rays of truth in the Vedic Literature, which amply prove the copious resources of our Vedic fathers in Geology, while their stray utterances in the matter, when duly connected, give us but a harmonious whole, and present their marvellous fertility of mind in unusual bright colours.

However, the geological knowledge of our Vedic fathers was, in the very nature of things, and as should be expected, very limited. Consequently, we have been able to obtain only a bird's eye-view of the geological past, as depicted by, or known to them, in accordance with their own notions, and the means and materials they then possessed.

In the last chapter, we noticed in brief, and gave a few details of, the Azoic, Palæozoic, Mesozoic, and the Cainozoic or the Tertiary Epoch. And it now seems necessary to make a rapid survey of the Glacial Period, and of the Quaternary Era which followed it. Now, viewing in the descending order the geological formations, we find the Tertiary Era followed by the Glacial Epoch (हिमयुग), which is remarkable no less for the thick sheets of ice covering the higher latitudes of Asia, Europe, and America, than for depopulating the immense tracts of the three continents, that before enjoyed genial climate and were teeming with population.

The causes of the Ice Age have not yet been ascertained, and many theories, therefore, in respect thereof, appear to have been in the field. Some hold that the intense Arctic conditions had arisen, owing to Geographical causes ; that is to say, owing to somewhat different distribution of land and sea-areas in Glacial times ; or to the general depression of land on one side, its elevation on the other, the indrift of Glacial currents, and the consequent formation of Ice-sheets ; or owing to a change in the position of the Poles.

While others, led probably by Dr. Croll, maintain that the Glacial conditions were due to certain *Astronomical* causes, *viz. the elongated ellipse.* For, at the present day, the ellipse formed by the Earth's orbit is approximately circular in form, but, as supposed by Dr. Croll, it was greatly elongated in the Pleistocene Period. This, it is argued, brings about the alternate glacial and genial periods, and explains the occurrence of warm inter-glacial Epochs, at long intervals, since, every 10,500 years, the physical conditions of the two hemispheres must have been gradually reversed, owing to the precession of the equinoxes. There is yet a third theory promulgated, and according to this, it is believed that, "even if the glacial conditions were due to astronomical causes, there were in reality no relatively warm inter-glacial periods, but merely local advances and retreats of the great Ice-sheet, owing to accidental, local, or annual variations in temperature, snow-fall, and the like". (*Vide* Lapworth's Text Book of Geology. *p.* 386. Ed. 1899).

Be that as it may, the last Glacial Epoch, known as the Great Ice Age, lasted for a considerable period (vide ante *pp.* 84, 85), and was followed by the Quaternary Era or the Recent period, when Ice-floods having subsided, order began to be restored. And such of the A′ryans as had left A′ryâvarta—their Home—and colonised the Arctic regions, but were compelled to abandon those genial tracts owing to the advent of the Great Ice Age, commenced returning Home to A′ryâvarta, or began selecting new settlements according to their choice and the surrounding circumstances.

I need hardly state, that our Vedic Fathers were aware of the Ice Age. But, this apppears to have been designated, in our Vedic Works, as the *Deluge* (प्रलयः, औघः). For, the oldest of these accounts has been incorporated in the Shatapatha[1] Brâhma*n*a (I. 8. 1. 1-10), and this only speaks of the *Deluge* (औघः). I have already made a passing reference to the Deluge which destroyed our magnificent and extensive colonies in the Arctic

---

1 In regard to the antiquity of this composition, Mr. Bal Gagadhar Tilak, the author of the *Orien* and the *Arctic Home in the Vedas* says, "the story of the Deluge is found in such an ancient book as the Shatapatha Bráhma*n*a, the date of which has now been ascertained to be *not later* than 2500 B. C., from the fact that it expressly assigns to the Krittikás, or the Pleiades, a position in the due East. It is evident, therefore, that the story of the Deluge is A′ryan in origin, and in that case, the Avestic and the Vedic account of the deluge must be traced to the same source." (Arctic Home in the Vedas *p.* 387.)

regions (*pp.* 61 @ 64, 72 @ 78, 82), and I would here only narrate the Fish-legend, in view of enabling the Reader to understand its origin and its connection with the Great Ice Age. I would first quote a few extracts from the original text in Sanskrit, for convenience and easy reference, and then give its translation as rendered into English by Muir :—

मनवे ह वै प्रातः । अवनेग्यमुदकमाजन्हुः...... । एवं तस्यावनेनिजानस्य मत्स्यः पाणीऽआपेदे ॥१॥ स हास्मै वाचमुवाद । बिभृहि मा पारयिष्यामि त्वेति । कस्मान्मा पारयिष्यसीति । औघ इमाः सर्वाः प्रजा निर्वोढा ततस्त्वा पारयितास्मीति । कथं ते भृतिरिति ॥ २ ॥ स होवाच यावद्वै क्षुल्लका भवामो बह्वी वै नस्तावद् नाष्ट्रा भवत्युत मत्स्य एव मत्स्यं गिलति । कुंभ्यां मा अग्रे बिभरासि। स यदा तामतिवर्धा अथ मा समुद्रमभ्यवहरासि। तर्हि वा अतिनाष्ट्रो भवितास्मीति ॥ ३ ॥ शश्वद्ध झष आस । स हि ज्येष्ठं वर्धतेऽथेति समां तदौघ आगन्ता तन्मा नावमुपकल्प्य उपासासै । स औघे उत्थिते नावमापद्यासैथतिततस्त्वा पारयितास्मीति ॥ ४ ॥ तमेवं भृत्वा समुद्रमभ्यवजहार । स यतिथिं तत्समां परिदिदेश ततिथिं समां नावमुपकल्प्योपासां चक्रे। स औघे उत्थिते नावमापेदे । तं स मत्स्य उपन्यापुप्लुवे । तस्य शृंगे नावः पाशं प्रतिमुमोच । तेन एतमुत्तरं गिरिमधिदुद्राव ॥ ५ ॥ (श० प० ब्रा० १. ८. १. १-१०).

"In the morning, they brought to Manu water for washing......As he was thus washing, a fish came into his hands, (which spoke to him) 'preserve me; I should save thee.' (Manu enquired), "From what wilt thou save me?" (The fish replied), 'A flood shall sweep away all these creatures; from it I will rescue thee.' (Manu asked), 'How (shall) thy preservation (be effected)?' The fish said, 'So long as we are small, we are in great peril. For, fish swallows fish.' Therefore, 'thou shalt preserve me first in a jar. When I grow too large for the jar, then thou shalt dig a trench and preserve me in that. When I grow too large for the trench, then thou shalt carry me away to the ocean. I shall then be beyond the reach of danger.' Straitway, he became a large fish; for, he waxes to the utmost. (He said), 'Now, in such and such year, (which, however, was not definitely specified), the flood will come. Thou shalt, therefore construct a ship, and resort to me. When the flood rises, thou shalt embark in the ship, and I shall deliver thee from it.' Having thus preserved the fish, Manu, carried him away to the sea. Then, in the same year, which the fish had enjoined, he constructed a ship and resorted to him. When the flood rose, Manu embarked in the ship. The fish swam towards him. He fastened the cable of the ship to the horn of the fish. By this means, he hastened (and got) to this Northern Mountain."

(Muir's Original Sanskrit Texts. Vol. I. *pp*. 182-3. Second Edition).

Now, the *Northern Mountain* in the above narrative evidently refers to the Snow-clad-Himâlayas ; and by the expression उत्तरं गिरिं, the commentator also understands the Himavat (हिमवत्) or the Himâlaya to be *the Mountain to the North of* A'ryâvarta and Bharata-Khanda or India. Besides, the phrase उत्तरं गिरिं shows but the past reminiscence of a very vivid re-collection of the great snowy moun ain, which in the hoary days of the Tertiary Period, *was observed* by our oldest ancestors *to the North of the region of the renowned Seven Rivers* (सप्त-सिंधवः), which was *the A'ryan cradle* and *the Birth place of the Primitive ancestors of* our Rig-Vedic Fore-Fathers, from where we had extended our conquests in all directions, and colonised the vast Arctic Regions in the Tertiary and Inter-Glacial Epochs ; but which, on account of the advent of the Great Ice Age, having become uninhabitable in the Pleistocene Period, we were compelled to return home, bag and baggage, by way of that everlasting snowy Peak of the highest mountain in the world—*the Himâlaya*—which has been *the Northern Boundary* of A'ryâvarta, whose *Majesty was ever all in all to us.*

यस्येमे हिमवन्तो महित्वा ...... आहुः ।
(ऋ० वे० १०-१२१-४.)

गिरयस्ते पर्वता हिमवन्तः .... ।
(अ० वे० १२१-११.)

This Himâlaya Mountain, before we colonised the Arctic and other regions, we used to behold in the North, while yet we were lying in our *cradle* of the *Seven Rivers* (सप्तसिंधवः); and this had saved us and given us shelter, when we were at the point of being buried beneath the thick sheets of Ice and Snow, that covered the Arctic tracts and the higher Northern latitudes in the Glacial Period.

I may here state, that the same story of the deluge, with a few variations and difference of names, appears in the mythologies of other A′ryan nations; and as it bears great resemblance to the Deluge story of Manu, described in the Shatapatha Brâhmaṇa, from which it seems to have apparently been borrowed by others, I venture to give the following extract from the History of Greece for the sake of comparison:—

"The enormous iniquity with which earth was contaminated—as Apollodorus says, by the then existing brazen race, or as others say, by the fifty monstrous sons of Lykaon—provoked Zeus to send a general deluge. An unremitting and terrible rain laid the whole of Greece under water, except the highest mountain-tops, whereon a few stragglers found refuge. Deukalion was saved in a chest or ark which he had been forewarned by his father Prometheus to construct. After floating for nine days on the water, he at length landed on the summit of Mount Parnesses. Zeus having sent Hermes to him, promising to grant whatever he asked, he prayed that men and companions might be sent to him in his solitude. Accordingly, Zeus directed both him and Pyrrha (his wife) to cast stones over their heads: those cast by Pyrrha became women, those by Deukalion Men. * * * * Deukalion on landing from the ark, sacrificed a grateful offering to Zeus Phyxios, or the God of escape; he also erected alters in Thessaly to the twelve great gods of Olympus." (Grote's History of Greece. Vol. I. Chapter 5).

A question, however, would naturally arise, that the catastrophe in the Shatapatha Brâhmaṇa refers to the Water-Deluge, as no mention is made of ice or snow in the whole narrative, and a such, this deluge might be some local flood consequent upon heavy showers of rain. But, there seems no reason to entertain any doubt about this, as the deluge in the Shatapatha Brâhmaṇa refers to the sweeping floods from the glaciated regions, and Manu appears to have been carried along these, in a ship guided by the fish, to the Northern Mountains or the Himâlayas. And in respect of this, Mr. Tilak also says that, "Nevertheless, it seems that the Indian story of deluge refers to the same catastrophe as is described in the Avesta, and not to any local deluge of water or rain. For, though the Shatapatha Brâhma*n*a mentions only a flood (*aughah*), the word *pralaya*, which Paṇini (VII-3-2) derives from *pralaya* (a deluge), signifies 'snow,' 'frost,' or 'ice,' in the later Sanskrit literature. This indicates that the connection of ice with the deluge was not originally unknown to the Indians, though in later times it seems to have been entirely overlooked." (Arctic Home in the Vedas. *p.* 387).

Moreover, the prophetic words in the Avesta (ante *pp.* 81, 82), like those of the Fish in the Shatapatha Brâhmaṇa (*pp.* 133, 137), corroborate and establish the fact that the Indo-A′ryan story of the Deluge refers to the same devastation by Ice and Snow-floods, during the Pleistocene Period, and not to any local floods of water or deluge caused by excessive rain.

Because, the close similarity in the Deluge-story as narrated in the Vedas and the Avesta, nay, the identity in the incident and even of the names of some of the chief characters therein, makes the evidence all the more relevant, and stronger still, in the matter. For instance, while on the one hand, the Fish warns Manu of the coming deluge in the Shatapatha Brâhmaṇa, and asks him to construct a ship for embarking therein, Ahuramazda on the other hand tells Yima, (supposed in the Avestic Scriptures to be the king of great wealth), about the coming winter-frost in a prophetic tone, after advising him to build a *Vara* (enclosure), for preserving all seeds therein. And I may here with advantage take the opportunity to observe, that Manu is also called Vivasvat (विवस्वत्) in the Rig-Veda,[1] and that the Yima of the Avesta is our Indian Yama. Besides, Yima's another appellation in the Avesta is also *Vivanhan*, which is but a corrupt form of Vaivasvata (वैवस्वत) appearing in the Rig-Veda,[2] where he is called by both the names, viz. Vaivasvata and Yama (वैवस्वतं...जनानां यमं राजानं...), and said to be the son of Vivasvan[3] (विवस्वंतं हुवे यः पिता ते) (यम).

---

1 यथा मनौ विवस्वति...। (R. V. VIII. 52. 1).

2 वैवस्वतं संगमनं जनानां यमं राजानं हविषा दुवस्य ।
(R. V. X. 14. 1).

3 अंगिरोभिरागहि यज्ञियेभिर्यम वैरूपैरिह मादयस्व ।
विवस्वंतं हुवे यः पिता तेऽस्मिन्यज्ञे बर्हिष्या निषद्य ॥
(R. V. X 14. 5).

The deluge, therefore, in the Shatapatha Brâhmaṇa is the same as the winter-frost of the Avesta, and both these evidently refer to the Pleistocene or Glacial Period, which lasted for a considerable time (*pp.* 84, 85), and was followed by the Quaternary Era.

Now, the Ice Age and the Quaternary Era appear to have been divided by our Vedic Geologists into *Krita* (कृत), *Tretâ* (त्रेता), *Dwâpara* (द्वापर), and *Kali* (कलि), of which, therefore, requisite details would be given presently, as they are very important and interesting. In the meanwhile, however, I shall give an extract from the Aitareya Brâhmaṇa, where, for the first time, mention is made of the four Yugas, and then notice in brief the nature of the changes, brought about by the advent of the Great Ice Age.

कलिः शयानो भवति
संजिहानस्तु द्वापरः ।
उत्तिष्ठंस्त्रेता भवति
कृतं संपद्यते चरन् ॥
(ऐतरेय ब्राह्मणम् ७-१५).

Muir translates the verse as follows :—

" A man while lying is the Kali ; moving himself, he is the Dvâpara ; rising, he is the Tretâ ; walking, he becomes the Krita." (O. S. Texts. Vol. I. *p.* 48. Second Edition).

Manu refers to this verse, and almost paraphrases it. But, in doing so, he happily gives us a direct clue to its real import, or possible original meaning. I shall, therefore, quote the same, along with its translation, as rendered into English by Muir, and explain the same later on.

कलिः प्रसुप्तो भवति स जाग्रद्द्वापरं युगम् ।
कर्मस्वभ्युद्यतस्त्रेता विचरंस्तु कृतं युगम् ॥
(मनुस्मृतिः ९. ३०२).

" While asleep, he is the Kali ; walking, he is the Dvâpara age ; intent upon action, he is the Tretâ ; moving about, he is the Krita. " (Muir's Original Sanskrit Texts. Vol. I. *p.* 49, Note. Second Edition).

We have seen that the Deluge or the so-called Ice-Age was a great catastrophe, as it destroyed our prosperous colonies and well populated tracts in the Arctic regions, and forced us to abandon them forever, and wander in search of lands free from the direful frost, as also ice and snow that buried beneath them enormous country, and extensive tracts of land. Such of us, however, as had great love for our *Mother Country-A′ryâvarta*, had never forgotten it. Nay, they had always considered it as their *Primitive cradle* and *the A′ryan Home*, and had ever remembered it with fondness that was simply unique (*pp.* 74, 75, 76). They therefore, naturally endeavoured to return home to A'ryâvarta, by way of the Himâlayas, while others sought refuge in places which either gave them shelter, or which they chose for themselves, having had due regard to the surrounding circumstances, and finally settled there.

This, in substance, appears to me to be the interpretation of the Sanskrit passage from the Aitareya Brâhma*n*a quoted above. However, referring to this Sanskrit-Text, Dr. Martin Haug takes a different but probably erroneous view. He first renders into English the passage as follows :-" The Kali is lying on the ground; the Dvâpara is hovering there; the Tretâ is getting up; but the Krita happens to walk (hither and thither)." He then says, " Sâya*n*a does not give any explanation of this important passage; where the names of the Yugas are mentioned for the first time. These four names are, as is well-known from other sources (see the Sanskrit Dictionary by Boehtlingk and Roth. S. V. *kali*, *dvâpara*, &c.) the names of dice, used at gambling. The meaning of this Gâthâ is, " There is every success to be hoped; for the unluckiest die—the Kali—is lying, two others are slowly moving and half fallen, but the luckiest, the Krita, is in full motion. The position of dice given here is indicatory of a fair chance of winning the game." (*Vide* Dr. Haug's translation of Ait. Br. Vol. II. *p.* 464. Note. Ed. 1863).

But, it seems that the passage in question refers to the picture, that presented itself after the advent of the Great Ice Age or the Pleistocene Epoch, and depicts in glowing colours the state of our Arctic Colonies subsequent to and during the catastrophe. I shall, therefore, revert to it, and endeavour to explain the same in detail, as the Vedic text has but an important bearing on the present chapter, from the stand-point of Geology.

I would first literally translate the passage thus :—" *Kali* is lying ; *Dvâpara* is abandoning ; *Tretá* is standing up ; and *Krita* is wandering. " Now, viewing in this light the original text, we shall try our best to explain what it means. After the hard winter made its appearance, and frost began to convert our once genial and habitable Arctic Colonies into thick sheets of ice, extending for miles together, our ancestors living in those northern regions, had a very hard and distressing time of it, and having been overtaken unawares, they were naturally at their wit's end, and had no remedy left except to abandon these colonies, and take shelter for protection from this impending danger, in such tracts as were free from the devastating ice. They were, therefore, compelled *to wander* (कृतं संपद्यते चरन्) take southern direction, and roam about night and day in quest of refuge and protection, till at last they passed the northern latitudes overrun by the deadly frost and ice.

As soon, however, as they stepped beyond the frozen zone, they *stood* and *stopped*, made a halt, and were intent upon some action (उत्तिष्ठंस्त्रेता भवति। *vide* Manu's explanation and paraphrase of this–कर्मस्वभ्युद्यतस्त्रेता–as also its English translation, ante *p.* 144), as they found time to breathe, evidently for the reason that they were beyond the reach of danger. But, although they had outlived the catastrophe, certainly not all but only a very few of them, they were, all of them, homeless, and could no more live in places where they had temporarily made a halt, only for breathing time. They, therefore, naturally considered that it was time for *quitting* these halting stations (संजिहानस्तु द्वापरः). Consequently, they *abandoned* these places where they had halted for a time, moved and marched onwards, in search of new homes and lands of their choice for settlement, till at last the junior branch–the European A'ryans who had forgotten A'ryàvarta, their *cradle*—settled in parts of Europe, now known by the name of Russia, Germany, Great

Britain, Italy, Greece, &c., and even distinguished one of these settlements by calling it *A′rya-Land* or Ireland; while the senior branch, the Irânian A′ryans settled in Irân (modern Persia), and the Indo-A′ryans—the oldest of the stock—hastened and returned home to A′ryàvarta, by way of the Himâlaya—called *the Northern Mountain* (उत्तरं गिरिं, ante *pp.* 63, 72 @ 77, 82 @ 84), and *lay and rested there* (कलिः शयानो भवति). This, therefore, was the beginning of the Quaternary Age, or *Kali Yuga.*

Now, the fact that when after the Great Deluge or the Arctic Catastrophe, caused by the incessant and heavy floods of Ice, our ancestor Manu thought of sailing to the South, from our Arctic Colonies, owing to the thick sheets of Snow and Ice that were fast covering the Arctic regions as also the higher latitudes, he had in his memory the aforesaid Himàlaya or the *Northern Mountain* (एतमुत्तरं गिरिम्), meaning thereby the mountain to the *north* of *A′ryâvarta,* where he and his fore-fathers, nay *the Primitive Ancestors,* were born, had lived, and *seen from their childhood* the great Mountain-Wall to the *north of the Land of their birth,* proves, beyond, all manner of doubt, that we (A′ryans) did not at first belong to the Arctic Regions, nor to Europe, nor to the Central Asiatic Plateaux, but had gone *there* from A′ryâvarta—*Our Hôme and Cradle*—for purposes of Colonisation and spreading our A′ryan Civilization over the Globe. For, had Manu never seen A′ryâvarta or the Great Himâlayan Mountain *to the north of the land* before, rather than have called *the highest wall* to the north "*the Northern Mountain*," he would certainly have termed it *the Southern Mountain*, as evidently it lay to the *South* of the Arctic Regions, or even Europe, and the Central Asiatic Plateaux, from where, it has been erroneously supposed by some, that we (Indo–A′ryans), *for the first* time, *immigrated into*, *not returned to*, the Land of the Renowned Seven Rivers. If, therefore, under the circumstances, *the stupendous wall* was still designated and called *the Northern Mountain*, it follows that it was *to the north of the Land of our birth*, and thus indicates a faint reminiscence and an ancient relic of an older order of things, or at any rate manifests a dim memory of our *Original Home* in A′ryâvarta, to the north of which the Himâlayan Mountain is situate, and owing to which fact, it was named *the Northern Mountain*, whose protection and refuge was sought by the Fish for the sake of Manu, by means of a ship, of which the cable was fastened to its horn (तस्य शृंगेनावः पाशं प्रतिमुमोच । तेन एतमुत्तरं गिरिमतिदुद्राव

शतपथब्राह्मणे १-८-१-५), when there was a Great Flood and the Ice had capped the northern latitudes.

In connection with this, and for corroboration of the fact of our *Cradle* having been in A′ryâvarta, I may with advantage quote here some noted researchers and great authors of the West, as that will serve as an unbiased testimony, and be considered as an independent source of evidence.

Says Curzon :—" From these considerations, it follows that there is not sufficient foundation for the hypothesis that the ancient A′ryans, Indians, or Hindus, entered India Proper from some external region. On the contrary, the facts above delineated point to the conclusion that the rise, progress advanced in the arts, and civilization of these remarkable people, are the growth of their own land and communicated to other nations..." (Journal Royal Asiatic Society of Great Britain and Ireland. Vol. XVI. May 1854. Part II. *p*. 199).

Höfer argues that since the most archaic forms of A′ryan speech are preserved in the R*i*g-Veda and the Avesta, the cradle of the A′ryans must have been in the region where Sanskrit and Zend were spoken. (Vide *The*

Origin of the A'ryas. By Isaac Taylor. *p.* 39. Second Edition). But, *A'ryávarta has only been the place* where Sanskrit language has been *spoken from time immemorial, as has been proved and said by Max Muller* (What can India teach us ? *pp.* 78, 79, 80. Ed. 1883); whereas, Roth says, that of the two sources, *viz.* the Vedic and the Avestic, " the Vedic has continued fuller, purer, and truer to its original character, while the other has become in many ways polluted "...... (Journal Germa Oriental Society 1848. *p.* 216

Cruiser, a French *Savant*, writes in unequival language that, " If there is a country on earth which can justly claim the honour of having been the cradle of the human race, or at least the scene of primitive civilization, the successive developments of which carried into all parts of the ancient world, and even beyond, the blessings of knowledge which is the second life of man, that country assuredly is India."

Moreover, M. Louis Jacolliot says :— "India is the world's cradle; thence it is, that the common mother in sending forth her children even to the utmost West, has, in unfading testimony of our origin, bequeathed us the legacy of her language, her laws, her *morale*, her literature, and her religion."

"Traversing Persia, Arabia, Egypt, and even forcing their way to the cold and cloudy North, far from the sunny soil of their birth; invain they may forget their point of departure, their skin may remain brown, or become white from contact with snows of the West......" (La Bible Dans L' Ind. *pp.* VII, VIII. Edition 1870).

In the same way, Count Bjornstjerna also says, "It is there (in A'ryâvarta) we must seek not only for the cradle of the Brâhmin religion, but for the cradle of the high civilization of the Hindus, which gradually extended itself in the West to Ethiopia, to Egypt, to Phænicia; in the East to Siam, to China, and to Japan; in the South to Ceylon, to Java, and to Sumatra; in the North to Persia, to Caldæa, and to Colchis, whence it came to Greece and to Rome, and at length to the remote abode of the Hyperboreans." (Theogony of the Hindus. *p.* 168).

Last but not least, Mr. Pococke writes follows :—"The mighty human tide that passed the barrier of the Punjâb, rolled onward towards its destined channel in Europe and in Asia, to fulfil its beneficent office in the moral fertilization of the world". (India in Greece. *p.* 26).

It is, however, not the province of this work to cite evidence and give details for proving the *A'ryan Cradle* in *A'ryâvarta*. I would, therefore, only recapitulate in brief some main points against the Arctic theory, the European hypothesis, and the Central Asian question, to avoid misunderstanding. The supposition that the tropics were too hot for life, even after the Polar regions and temperate zones were inhabited, "is open to question on physical grounds, and appears contradicted by the similarity of Silurian fossils in the Southern hemisphere to those in the northern." (*Vide* Manual of the Geology of India. By Messrs. Medlicott and Blanford. *p*. XXII). Besides, "there are very curious indications of a low temperature having prevailed in the Indian area, in very ancient epochs." (Manual of Indian Geology. *p*. XXII).

Moreover, the discovery of the fossils belonging to *the Vindhyan* or Algonkian Era, in *the Salt Range of the Punjab*, indicates that A'ryâvarta or *the Land* of the *Seven Rivers* (सप्तसिंधवः) having sufficiently cooled down, had enjoyed a low temperature and mild climate, even during *Pre-Cambrian* and subsequent Epochs. Since, life in A'ryâvarta has been proved by geological evidence to be older than even *the Lowest Cambrian Age*, as

Dr. Noetling had declared to say that, a series of strata containing fossils called *Olenellus* and various remains of Annelida, &c., was found in North-West India, and in *the Salt Range cf the Punjab*, and that they were *really of older age than the Lowest Cambrian*, that is, of the *Purána, Vindhyan, Pre-Cambrian*, or Algonkian Era. (*Vide* ante *pp.* 102,114; *The Student's Lyell.* Edited by Judd. *p.* 438. 1907 ; The Imperial Gazetteer and the Indian Empire. *Vol.* I. *p.* 55, Ed. 1907). While, in the Arctic regions, vitality does not seem to have been even of the Ordovician Era, much less of the Cambrian Period, as geologists appear to have noted life in the Arctic regions only from *the Silurian Epoch.* (*Vide* Intermediate Text-Book of Geology. By Lapworth. *p.* 228, Edition 1899. Dr. Dana's Manual of Geology. *p.* 206. Edition 1863).

Further, the theory that the A′ryans *migrated* from North to South, and as such were *immigrants* in A′ryâvarta or *the Land of the Seven Rivers*, does not hold water, as "the cases are not sufficiently numerous to indicate any law of migration from North to South" (*vide* Medlicott and Blanford's Manual of Geology. *p.* xx), and 'the tendency to survival in India is not universal.' (Manual of Geology. *p.* Lxx).

In the light of these facts, as also the Vedic, Avestic, Foreign, and Geological evidence hereinbefore cited (*pp.* 34 @ 76, 77 @ 82, 83, 141 @154), we can fairly say that, it was from A'ryâvarta or the Land of the renowned Seven Rivers —*our Home and Cradle*—that we had gone to the Arctic Regions, where our R*i*g-Vedic Fore-fathers and their hoary ancestors had established stupendous fabric of our *Colonial Empire.* While enjoying there the genial climate of the place (ante *p.* 63), signs of destructive frost became visible all of a sudden, and in no time, there were seen devastating floods of thick sheets of Snow and Ice everywhere. Evidently, the Great Ice Age had made its appearance, and it had affected the higher latitudes alike of the Old and the New world. This, therefore, compelled Manu, the leader of our Colonies, as also other Colonists to abandon the once genial regions. Manu, accordingly, endeavoured to return *Home* to A'ryàvarta, while others effected their retreat to such places as could afford them shelter from the ravages of the Ice-floods, or as they were able to choose under the circumstances (*vide* ante *pp.* 135, 138, 144, 145, 148).

Now, just as in the Vedas, so also in the Avesta, we have unassailable evidence in respect of the advent of the Great Ice Age called the Glacial Period, to which, therefere, reference has already been made from time to time (*vide* ante *pp*. 35, 63, 72, 73, 76, 77, 81, 82). Our Vedic Geologists, however, had divided this Glacial Epoch of enormous length (*vide* above *pp* 84, 85), into three Yugas, *viz. Krita* (कृतं), *Tretâ* (त्रेता), and *Dvâpara* (द्वापरः); while *Kali* (कलिः) was according to their notion, the beginning of the Quaternary Age. All this, therefore, seems to have been very well described and scrupulously observed by our Vedic R*i*shis, which no ancient nation whatever had done, with the eye of an expert, the zeal of the Geologist, and the accuracy of a careful observer, as due regard appears to have been paid to the functions of each Yuga, of which, by the bye, the requisite details have been given before (*vide* ante *pp*. 144, 145, 146 @ 148), and in mentioning which, our Vedic R*i*shis have certainly shown themselves to be immeasurably in advance of the times.

In fact, just as the Great Ice Age finds place in the Science of Geology as cultivated in the Occident, the aforesaid four Yugas form but one important Episode in the Vedic

Geology of the Orient. The Reader, therefore, will have easily perceived, that our Vedic fore-fathers having seen the Great Ice Age, they were good enough to kindly give to their progeny and to the world the true picture and lively account of what they themselves had seen, heard, or known by tradition, and have thereby placed them under deep obligation. Thus, the afore-said passage from the Aitareya Brâhma*n*a, quoted above on page 143, presents to us a vivid sketch of the Glacial Era, and of the commencement of the Quaternary or Recent Period in its true colours, though the language used therein seems to us, at this distant date, enigmatical, and the sense abstruse. However, the description though brief, is yet terse and graphic, and brings vividly to our mind the chain of events as had then occurred. (*Vide* ante *pp.* 144 @ 149).

ॐतत्सद्ब्रह्मार्पणमस्तु ।

INDEX.

A.

Advent of Man p. 109, 128, 130.
–of the Great Ice Age 63.
Age of Mammalia. 22.
–of man 22.
–of reptiles 22.
Agni–the Fire-God 67.
Agni Purâṇa 88.
–Geological references in-10, 88.
Ahi 49. (vide Serpent).
Ahurmazda 78, 80, 82, 142.
Aitareya Brâhmaṇa 74, 99, 143, 145, 157.
Algæ 110, 111.
Algonkian 153, 154.
America 133.
American Geologists 84.
–Their opinion about the close of the Glacial Epoch 84.
Amphibians 115.
Amphibious life-types 115.
Annelida 114, 154.
Anthropoid Primate 89, 90.
Anticipated hypothesis of our Vedic Fathers in Geology 95.
Antiquity of the Vedas from the Geological point of view *p*. IX, 25, 28, 64, 67, 69, 72, 83.
Apollodorus 139.
Aquatic beings 96.
Aqueous vapour 95.
Arabia 152.
Arabian writers on Geology 97.
Aranyaka–Taittirîya 14.
Arctic *p*. III.
Arctic catastrophe 148.
Arctic colonies 81, 82, 83, 146, 148.
Arctic conditions 133.
Arctic regions 34, 35, 57, 62, 68, 69, 70, 72, 73, 75, 76, 77, 80, 82, 135, 139, 144, 148, 149, 154, 155.
–become unsuited for habi-

tation of man, beast, and plant 72.
-climate in the Arctic regions milder 63, 72, 73, 80.
-and genial 82, 83.
-colonised 80.
Arctic theory 153.
A'rjikiya- (A vedic river) Vipâsh or Biaj 56.
A'ryans 39, 75, 76, 135, 148, 150, 154.
Indo-A'ryans 148.
Iranion A'ryans 148.
Primitive A'ryans 59, 60, 61.
Sacrifice loving A'ryans 58.
A'ryan- Civilisation 29, 61, 149.
A'ryan cradle 34, 53, 77, 98, 138, 149, 150.
A'ryan- European 147.
A'ryan home 99, 144.
A'ryan Indian (Indo-A'ryan) 36, 66, 78, 79, 81, 82.
A'rya- Land (Corrupt form Ireland) 148.
A'ryan nations 139.
A'ryan race 73.
A'rya-worshippers 40.
A'rya'varta *p*. III, 34, 35, 36, 37, 38, 39, 43, 46, 47, 52, 67, 68, 69, 74, 75, 77, 78, 79, 80, 81, 82, 98, 99, 101, 102, 135, 138, 144, 145, 147, 148, 149, 150, 151, 152, 153, 154, 155.
-colonies of-60.
-Emigration from-34.
-the Mother-Country- 34, 79, 81.
-the Mother- land. 35, 76.
-the Origin of Indra and Soma in-45, 46.
-the Sacrificial region 36.
-the *scene* of creation 35, 76.
A'rya'vartic Home 34.
Ashvins 66.
Asia 133, 152.
Asiatic Plateaux 62.
Central, Asiatic Plateaux 148, 149.
Asikni (A Vedic river, now known by the name of Chinâb or Chandrabhâgâ) 78.
Assyrians 98.
Asuras 76.
Atharva Veda 69, 70, 73, 74.
expression of fear on account of the extreme length of nights in the Atharva Veda 69.
Aughah 136, 141.
Autochthonous in India,
-we Indo-Aryans 34.
-the Primitive Ancestors of our Rig-Vedic Forefathers 36, 57, 82, 149.
A'vestic evidence in support of the Aryan-cradle in A'ryâvarta 155.

Azoic Age 110.
-Epoch 21, 104, 105, 107, 115, 128.
-Period 128.

B.

Babylonians 98.
Beeas (Bias or the river Bejah). *vide* Vipâsh, the Vedic name of modern Bias 55, 56.
Benefactors of Mankind-the Vedic A'ryan Fore-fathers *p*. I.
Bha'rata Khanda 138.
Bhagavata Purâṇa-Geological references in the-9.
Bharata-Varsha 34, 75.
Bha'rati'ya Sâmrâjya-or the History of the Hindu Empire *p*. III.
Bha'va'rtha D*i*pikâ. or Jnaneshvarî 123.
Birth-place-of the Primitive Ancestors. 138.
Bjornstjerna-Count.
His views about the A'ryan cradle being A'ryâvarta 152.
Blanford 153.
Bloomfield-Professor-on the antiquity of the Vedas 26, 30.
Boar (वराह) 18, 88, 116.
-representing the mammalian life-type 14.
-the mammalian order represented by *Boar*, as per view of the Vedic and Purâṇic Geologists 116,128.
Boehtlingk 145.
Boyle 97.
Bra'hmanam 26.
-the approximate date of the Brâhmaṇas 26, 27.
-the Aitareya-74, 99, 143, 145, 157.
-Kaushitaki 74, 99.
-Shatapatha-27, 77, 114, 135, 136.
-Taittir*i*ya 13, 14, 16, 17, 18, 19, 20, 27, 28, 70, 89, 105, 106, 116, 119, 120, 129.
Brahmanical Books 44.
Bra'hmans 36.
-the worshippers of Indra 58, 59.
-the Sacrifice-loving A'ryans 58.
-those who know the Brahma Mantra, or the land of the Lord 58.
Brahmarshi Desha 38.
Brahma'varta 35, 38
Brihaspati 27, 28.
Buddhist learning *p*. V.
Buffon 97.
Buhler-Dr.-on the antiquity of the Vedas 26.
Bukka-King of the Empire of Vizayânagar 123.

C.

Cainozoic Epoch 103, 112, 113, 133.
-Era 108, 116.
-the age of mammals 22, 117.
-or the Tertiary times 11.
Calcareous algâe 111.
Caldæa 152.
Calloway-Mr. Charles 112.
Cambrian Epoch 85.
-Period 154.
-the Lowest Cambrian Age 153, 154.
Canada-Eozoon or the dawn animal of- 129.
Causes of the absence of the Vedic and Paurâṇic Geological works *p.* V, VIII, 6, 7.
Cave or चिलं- Waters imprisoned in the-by Vr*i*tra 51.
Central Asia 57.
-Asiatic Plateaux 77, 148, 149.
Ceylon-the extension of A′ryâvarta in-152.
Chhandogyopanishad 124.
Chandrabha′ga′ or the Vedic river Asikni-78.
-the modern Chinâb 78.
Chaotic condition of Earth 8, 128.
-Ocean 94.
China′b 78.
China 152.
Chitra′vasu-the night 67.
Church Hill 129.
Circum Polar Regions 34.
-A′ryan colonisation of the-34.
Clodd (*vide* Edward Clodd) 33.
Close of the Tertiary Epoch 84.
-Time that elapsed since the-84.
Cloud (मेघ) 49.
Cloud-Serpent 46, 49. (*vide* Vr*i*tra).
Clue to the A′ryan Cradle and Original Home 53.
Coal Era 110.
-Measures 108.
-Bone-clad fishes of the-108.
Coalchis 152.
Colebrooke-His opinion in respect of the antiquity of the Puraṇas 10.
Colonial Empire of A′ryâvarta, and its stupendous fabric erected by our R*i*g-Vedic Fore-fathers and their hoary ancestors 155.
Colonization of the Vedic A′ryans in distant climes after leaving A′ryâvarta-their Mother Country 34.
Colony *p.* III, 34.
-from A′ryâ-varta, in parts of Asia 34, Africa 34, Arctic regions 34, 146,

Circum Polar-regions 34, Europe 34, Ame-rica 34, and every where 61.

Condensation of the aqueous vapour causing the fall of rain and giving rise to the first Thermal ocean 95.

Connecting link between man and beast, or the intermediate link between the brute intelligence and the reasoning powers of man 88, 89.

Cosmos 15. Evolution of–17.

Cosmogony *p*. VI.

Cosmology *p*. VI.

Cows- Meaning light and water–57.

Cradle (of the) A′ryans 38, 153, (*vide* A′ryâvarta).
–of the Brahman religion 152.
–of the high civilization of the Hindus 152.
–of the human race 151.
–Primitive Cradle and the A′ryan Home- *viz*. A′ryâvarta 144.
–of the world 151.

Creation- scene of–being A′ryâvarta–38, 75.

Creative energy dormant in the fluid 87.

Creator or the Prajâpati.
–of Palâeozoic vegetation and the fish 14.
–of the Mesozoic tortoise 14.
–of the Tertiary mammalia 14.

Crocodile 115.

Croll–Dr.–84, 134. His Work *Climate and Time* & *Climate and Cosmology* 84.

Crowning piece of Creation–the Man 88, 108, 128, 131.

Cruiser–His views about the Cradle of the human race being India 151.

Crust of the Earth solid 104, 127.

Cryptogams 110.

Crystalline 96.

D.

Dana, Dr. James. D.
–the American Geologist–16, 63, 110, 154.
–His views about the scope of Geology 5.

D′asa–the foe of A′rya–66.

Dasyus–the Vritra-serpents–58.

David Page. Dr. see Page.

Dawn 39, 40, 49, 52, 53, 54, 55, 56, 64, 65, 68.

Dawns-the everlasting-34,64, 65, 68, 72.
–continuous 63.
–long 69.

Day–phenomena of long–34, 64, 65, 69, 72, 83.
–Six monthly 64, 70, 80.

Dawn animal 129.
Dawson-Dr. 129.
Deluge 76, 135, 139, 140, 141, 142, 143, 144, 148.
-of Noah 93.
-Story of the-77, 139, 141.
-Vedic and Avestic account of the-135.
Destruction of Sanskrit Literature and of the Hindu well-stocked libraries *p.* V,VIII, 7.
Deukalion 140.
Devonion flora 110.
Dixit S. B.-on the antiquity of the Vedas 26, 27.
Donati 97.
Drishadvati-a Vedic river-35, 38.
Dwa'para (Yuga) 143, 144, 145, 146, 156.
Dwarf-Incarnation 88.
-Man or Wâmana 88, 112, 116.
Dyaus-the daughter of the sky 54.

E.

Earth-Chaotic state of the-8, 17.
-configuration of the-14.
-crust of the-*p.* IV.
-egg-form of the-9.
-gaseous state of the-13, 14, 15, 17, 19, 20, 104, 106. 121, 125, 126.
-liquid state of the-8, 9, 10, 13, 15, 16, 17, 18, 19, 20, 105, 106, 107, 120, 121, 126, 127, 128.
-solid state of the-13, 14, 15, 16, 18, 20, 21, 96, 10[illegible], 105, 106, 120, 121, 122, 126, 127, 128.
-mass of intense heat 10, 13, 17.
-primary condition of the-17, 104.
-vitality on-8, 9, 10, 13, 14, 15, 16, 18, 112.
Edward Clodd-President-Folk Lore Society 33.
-His Work The *Story of the Primitive Man* 33.
Egypt 152.
Egyptians 98.
Elphinstone.-*p* II.
-His History of India *p.* II, 37.
-His views about the Hindoo Cradle 37.
Emigration from, not immigration into A'ryâvarta. 36, 57, 62, 149.
England-Geologists of-97.
Eozoon-an organic structure-or the Dawn animal of Canada 129.
Episode of the Glacial and Quaternary Period *p.* IX, 132, 133, 135, 137, 139, 141,143, 145, 147,149-157.

Ernst Hæckel.
-Professor 32.
Ethiopia 152.
Europe 77, 91, 92, 133, 147, 148, 152.
-Northern 57.
-thinkers of modern-91.
European hypothesis 153.
Evolution of Cosmos 15, 17.
-of Earth 15.
-of vital gradation 15, 16, 87, 88.
Evolutionists-the Hindus-*par* excellence 91.

F.

Fargard-Zend or Pârsee Scripture 78.
First heroic deed of Indra, *viz*. the slaying of Vritra-Serpent 52, 53.
-observation by the Rig-Vedic Fathers and their Primitive Ancestors, of the fall of rain or the dropping of waters from the clouds 49, 50, 51, 52, 57.
-observation of the Dawn 52.
-observation of the morning light 57.
-observation of the sun 53, 57. (*vide* the second and the third deed of heroism of Indra).
Fish 14, 77, 88, 113, 117, 131, 141, 142, 149.
-sailing to the south 77.
-legend of the-136.
Forefathers of yore. *p*. II.
Foreign evidence in support of the A'ryan cradle in A'ryâvarta 155.
Foreign origin of the Indo-A'ryans or Hindus disputed and refuted 37.
Forests of A'ryâvarta 76.
Fossil-fish 129.
-plant 110.
Fossiliferous Conditions 94.
Fossils *p*. IV, 96, 102.
Fracastoro-97.
France-Geologists of 97.
Frederick Wright-Dr. 32. His proof of the Tertiary Man 32. His Work *Man and the Glacial Period*.
Fuchsel 97.
Further India—Relics of Man found in—33.

G.

Ganges -the river—76, 78.
Gaseous state of the Earth, 13, 14, 15, 17, 19, 20, 86, 105, 106, 119, 120, 121 122, 125, 126, 127.
G'atha's-Pârsi Scriptures 79.
Geikie-Professor 84.
Geological knowledge of the Vedic times *p*. IV, 4, 17, 25, 26, 91, 132, 157.

–literature Indian 6.
Geological treatises, books, or works–cause of their absence *p.* V, VIII, 7.
Geological evidence in support of the A'ryan cradle in A'ryâvarta 155.
Geological Science–state of the.–during the Vedic period–4, 108, 109, 110.
Geological succession 25.
Geology–Definition of–1.
–as given by Lyell 1.
–by Judd 2
–by Lapworth 2, 3.
–by Dana 3..
–discoveries in–by Purânic and Vedic Geologists–8, 9, 10, 100, 104, 105, 106, 107, 108, 109, 129.
–germs of-in the Vedic and Purânic literature. *p.* IV, V.
–scope and province of–1, 2, 3, 4.
–Vedic Fathers of–1, 91.
Germany 147.
–Geologists of–97.
Girvanella 111.
Glacial Age 84.
–conditions 134.
–currents 133.
–Epoch 82,84, 85, 133,156.
–Era 157.
–Period 81, 82, 84, 132, 133, 134, 139, 143, 156.
–times 133.
Gneiss 96, 120.
–formation of–14.
–origin of–16.
God of escape 140.
Gods 2:, 76, 113.
–portions of human nature 112.
–superhuman beings 113.
Granite 95, 96, 120.
Granite Crust 16, 21, 121.
Gravel 120.
–formation of–14.
–origin of–16.
Great Britain 147–148.
Great Deluge 75.
Great Ice Age 134, 136, 155.
–Advent of the–63, 72, 80, 135.
Greece 98, 139, 148, 152.
–Grote's History of–140.
Griffith R. T. H. 55, 62, 66, 100, 112.
Grote–Historian of Greece 140.
Guiding Mark–the Himâlaya or *the Northern Mountain.* 81.

H.

Hapta Hendu in Zend being the Sapta Sindhus–(सप्तसिंधवः) of Sanskrit 78.
Harahwaiti or the river Sarasvatí 78.
Haugh-Dr. Martin-His Essays on the Pârsi Religion 80, 82, 145.

Herbs (ओषधय:) 16, 18, 22, 109, 112, 113, 128, 130.
Hermes 140.
Hima'laya-the northern boundary-35, 76, 138, 141, 148,
-the chain of-37, 41,
-the Mountain 39, 41, 73, 74-75, 76, 77, 81, 138, 139, 145, 148, 149.
Himavat 138.
Hindus 91, 150.
-the Darwinians 91.
-the Evolutionists 91.
-the Philosophers 122.
-the Spinozites 91.
Hindu Kusha-the Mountain. 62.
Hindu life-Religion pervading in the-14, 107.
History of the Hindu Empire *p.* III.
-of Greece 139, 140.
Hofer 150.
Hoary traditions received from our progenitors 61.
Homa or Hoama of Irân being the Soma (सोम) of A'ryâvarta 44.
Home and Cradle *viz.* A'ryâvarta 61, 68, (*vide* Aryâvarta or the Land of the Seven Rivers).
Hooke 97.
Hunter-Dr. Sir William. -His reference to the age of Manu, our Law-giver 8.
Hutton-Dr. James-of Scotland-. VI, 95, 124.
Huxleys 91.
Hyperboreans 152.
Hypothesis in respect of the A'ryan Cradle, *viz.*
-the Arctic Theory 153.
-the European Hypothesis 153.
-the Central Asian question 153.

I.

Ice 81, 140, 141, 147, 148, 150, 155.
Ice Age 33, 133, 135, 144.
-the Great Ice Age 26, 31, 33, 63, 72, 83, 134, 138, 143, 155, 156, 157.
Ice-caps 35.
Ice-floods 81, 134, 148, 150, 155.
Ideal-love for A'ryâvarta -our Mother-Country 73.
Igneous fusion-the Earth in the state of-19, 93, 95, 105, 106, 107, 119, 122.
Imperial Gazetteer 102, 154.
Incarnation 130.
-of Boar 88, 112, 116, 131.
-of Dwarf Man 88, 112, 116, 131.
-of Fish 88, 111, 130.
-of God 116.
-of Man-Lion 88, 112, 131.
-of Râma 116.
-of Tortoise 88, 112, 130.

India *p*. II, 75, 98, 138, 151.
-Proper 150.
Indian area-low temperature in the-153.
Indian Magazine and Review 91.
Indians 141, 150.
Indian A'ryans or Indo-A'ryans 36, 66, 78, 79, 81, 82, 148, 149.
-autochthonous in India 36, 57, 149.
Indo-A'ryan life-religion pervading in-14, 107.
Indra 39, 40, 42, 43, 44, 45, 46, 47, 48, 49, 50, 51, 52, 54, 55, 57, 59, 60, 62, 66, 75, 104, 127.
-Antiquity of-43.
Indus-the river-40, 51, 52, 60, 61, 76, 78.
Inter-Glacial Epochs 84, 134, 138.
Ira'n 44, 79, 80, 81, 83, 148.
Ira'nians (Perso A'ryans) 78, 81, 148.
Ira'vati—the Vedic river Parushni or modern Râvi 78.
Ireland-the A'rya land 148, (*vide* Max Muller's *Science of Language*. vol. I. *p*. 274. Fifth Edition. 1866.
Italy —Geologists of—97. Modern Cradle of civilisation 98.

J.

Jacobi-Dr.-of Bonn, on the antiquity of the Vedas 26.
Jacolliot M. Louis 151.
Jamna' 78. (*Vide* Yamunâ the river).
Japa'n 152 (Hindu colonies in).
Ja'va'-Colonised by the Hindus 152.
Jna'neshvara-the great Mahârâshtra poet, saint, and philosopher 123.
Jna'neshvari' 123.
Jones-Sir William *p*. II.
Journal of the Bombay Branch-Royal Asiatic Society 27.
-of Great Britain and Ireland 150.
-of Oriental Society of Germany 151.
Judd J. W. Professor of Geology. *p*. VI, VII, 2, 85, 97, 102, 110, 111, 113, 114, 115, 117, 154.
Jupiter 27 (*Vide* Brihaspati).

K.

Kali 143, 144, 145, 146, 156.
Ka'shmir—the Himâlayan high land 45.
Keane-Ethnology of-32.
-His Work-*Man Past and Present* 32.

Krita-Yuga 143, 144, 145, 146.
Krittikas 27, 135. (*Vide* Pleiades).
Ketkar V. B.–on the antiquity of the Vedas 26.
Kulluka–the great commentator of Manu 37.
Kurukshetra 42.
Kaushitaki Bra'hmana 99.

L.

La Bible Dans L' Inde, or *The Bible in India*. A work of M. Louis Jacolliot 152.
Land–appearance and rise of –(and mountains) above waters 16, 21, 121, 122.
Land–of our birth (that is, of the A'ryans) 149.
–of the brave and the pious 75.
–of choice regions 76.
–of commerce and trade 75.
–of equal days and nights 76.
–of heroism and enterprise 75.
–of the Heaven on Earth 76.
–of medicinal herbs and plants 76.
–of *Patriotism* and *self-Sacrifice* 76.
–of *Sacred* pleasures 75.
–of Sacrifices 76.
–of the *Scene* of *Creation* 76.
–of the *Scene* where life had first commenced 75.
–of Science and Art 76.
–of the Seven Rivers 75. (Vide the Seven Rivers).
–of Six seasons 76.
–of valour and renown 76.
–of virtue and greatness 76.
–of virtue and kindness 76.
Land–plants 110.
Lapworth-Charles–2, 126, 129, 134, 154.
–His views about the scope of Geology 2, 3.
Laurentian rocks 129.
Lee–J. E. 129.
Lehman 92, 93, 94.
Leibnitz 92, 93, 125.
Libraries–Destruction of p. V, VIII, 7.
Life in the fluid of the deep dormant 18.
–on Earth 8, 9, 10, 16, 18, 21, 22, 88, 101, 128.
–in weeds, or herbs 22, 87, 88, 108, 110, 111, 112, 113, 128, 130.
–in lotus 87, 88.
–in moss 87.
–in plants 87.
–in trilobites 87, 108, 111, 128.

-in fishes 87, 88, 128.
-in reptiles 87, 88.
-in quadrupeds 87, 88, 89, 90.
-in man 87, 88, 128.
Life types-various kinds of-mentioned in the Sanskrit literature *p.* IV, 88, 111, 112, 114, 116, 130, 131.
Lightning or Indra's thunderbolt 48.
Lionardo da Vinci 97.
Liquid state of Earth 8, 9, 10, 13, 14, 15, 16, 17, 18, 19, 20, 87, 95, 104, 105, 106, 107, 119, 120, 121, 126, 127, 128.
Literature Vedic, Purânic, Classical, *p.* IV.
-of India, the most ancient 19, 25, 64, 83.
Logan-Sir W. 129.
Lost types of the anthropoid primates, 89.
Lowest Cambrian Age, 102, 153, 154.
Lozzaro 97.
Ludlow formations-upper-129.
Lykaon 139.

M.

Mc-Gee 85.
Ma'dhava Sa'yana-the Vedic Exegetist and the celebrated Prime-Minister of king Bukka 123.
Magadha Kingdom-the Centre and Head quarters of Buddhist learning *p.* V.
Maha' Bha'rata 37, 38, 39, 41, 45, 71, 88.
-Antiquity of-10, 11.
-Geological references in-8, 11, 23.
Mahomedan conquest and the destruction of Sanskrit literature *p.* V, VIII, 7.
Ma'lati Ma'dhava 90.
Mammalia-Tertiary-10.
Mammalian life types 14, 21, 128.
-order 109, 115, 116.
Mammals-age of-22.
Man 16.
-the advent of-21, 49.
-the age of-22.
-the crowning piece of creation 10, 87, 131.
Mankind-benefactors of-*p.* I.
Man-lion 88, 89, 90, 116.
Manu-the A'ryan Progenitor and ancestor 76, 77, 136, 137, 139, 140, 142, 148, 149, 155.
Manu the Law-Giver, 8, 35, 36, 37, 71, 144, 147.
-His Age 8.
-the importance and the universal extent of his Code 9.
-Geological references in his Code 8, 9, 144, 147.

Marine plants 110.
Max Muller *p.* I, II, V.—His views on the antiquity of the Vedas 19, 26, and their originality 64.
Mayne, John D.–His opinion in respect of the importance of the Code of Manu 9.
Medlicott 102, 153.
Megha 49.
Mesozoic Age (of Reptiles), or Secondary Epoch 22, 117.
-Epoch *p.* IV, 103, 108, 111, 114, 115.
-Era 112.
-Period 22.
-Reptile 10, 22.
-Tortoise 10, 14.
Metaphysics—evolution of Cosmos in–17.
Mica 96, 120.
Michell 97.
Mineral mutations 17.
Miocene deposits 31, 33.
-Era 11.
-Period 32.
-Upper Miocene deposits 33.
Miocene-Man–relics of–33.
Mlechha or Foreigner 36.
Mlechha देश, or country of the foreigners 36.
Molten or liquid state of Earth 8, 9, 10, 13, 14, 15, 16, 17, 12,8 19, 20, 87. 95, 104, 105, 106, 107, 119, 120, 121, 122, 127, 128.
Monier Williams–Sir
Moro 97.
Mother country–A'ryâvarta 63, 68, 72, 73, 76, 78, 79, 81, 82, 144.
Mother-Land-A'ryâ varta 62 65, 73.
Mountains out of water or liquid 96, 104, 121, 122.
Mountain Wall 148.
Moujavat Mountain 41.
Mriga 27. (*vide* Orion).
Muir 44, 119, 136, 143, 144.
Muir's Original Sanskrit Texts 35, 44, 106, 119, 138. 143, 144.
Munjavat mountain 41.

N.

Na'land library–destruction of–*p.* V, 7.
-the great seat of learning in India, *p.* V.
Narbada' valley 31.
-valley gravels of–containing *Shiwalic* fossils 31.
Natural religion–Fathers of *p.* I.
Nature–the phenomena of–*viz.*
(1) the fall of rain,
(2) the splendour of the Dawn, &
(3) the rise of the sun 49, 52. 53.

Neptunian theory 94.
New world 63, 155.
Niagara falls 84.
Night invoked and requested to become fordable 67.
-long-64, 65, 66, 67, 69, 70, 72, 83.
-six monthly-64, 70, 80.
Nile 98.
Noah Deluge of 93.
Noetling Dr. 32.
-discovery of fossils by-of Pre-cambrian Age-102, 111, 114, 129, 154.
Non Sacrificers 40, 42.
Northern Boundary of A'ryâvarta-the Himâlaya-138.
Northern India-most ancient life-types.-found in-101.
-origin of life in Northern India, or the region of the river Sarasvatî 101.
-the *scene* of *Primitive vitality*-A'ryâvarta-102.
Northern Mountain 76, 77, 137, 138, 141, 148, 149.
North-West India 102, 111, 129.
Nrisimha Purva Tâpani-Upanishad 89, 116.
Nycthemeron -one continuous day and night, making together a time of twenty-four hours 71.

O.

Occident 156.
-The Great Ice Age in the Geology of the-156.
Oldest Ancestors-*viz.*
-the ancient fore-fathers (of our R*i*g-Vedic R*i*shis) that belonged to the Tertiary Period *p.* I.
-the most ancient Poets of our thoughts *p.* I.
-the benefactors of Mankind *p.* I.
-the fathers of the most natural of Natural religions *p.* I.
-the framers of the most wonderful language the Sanskrit *p.* I.
-the makers of the most transparent of mythologies *p.* I.
-the most genuine inventors of very subtle philosophies *p.* I.
-the first givers of the most elaborate laws *p.* I.
-the first workers in the construction of our fundamental concepts *p.* I.
-the oldest ancestors that framed our first words *p.* I. (*vide* Primitive Ancestors).
Old world 63, 155.
Olenellus 102, 114, 154.
Olympus 140.
Oolite Epoch-108.
-reptiles of the-108.

Ordovician Era 154.
-Flora (that is, Siluro-Cambrian flora) 110.
Orient 157.
-the Great Ice Age in the Science of Geology of the-157.
Original Home and the A'ryan Cradle 53, 75.
-Home 63, 69, 149.
Origin of life 98.
-in the region of the river Sarasvatî 99, 100, 101.
Orion or Mriga 27, 28, 30.

P.

Page-Dr. David-110, 124, 125, 129.
Palæontological details 121.
-records *p.* VI.
Palæontology 103, 108, 128.
Palæozoic Epoch *p.* IV, 22, 103, 115, 128, 133.
-Era 21, 108, 111, 113.
-Fauna and Flora 117.
-Fish 14, 111.
-Herbs 108, 111.
-Lotus or weeds 14.
-Period 112, 117.
-Plants 108, 111.
-Trilobites 108, 111.
-Vegetation 14.
-Vitality 11, 109, 111.
-Weeds 14, 108.
Palissy 97.
Pandit S. *p.* 55, 100.
Pa'nini 141.
Parjanya (Rain) 39, 41.
Parnassus-Mount 140.
Parushni-the Vedic river -known also by the name of Irâvatî or Râvî-(the Modern name) 78.
Patriotism-land of-76.
Penance (Prâyaschitta) which even the Gods had to perform, whenever the Sun did not rise, as expected 68.
Persia' 148, 152.
Phænicia' 152.
Planet (the Earth) 109.
-Its origin 118.
-Its first gaseous state 13, 14, 15, 17, 19, 20, 86, 105, 106, 107, 119, 120, 121, 122, 125, 126, 127.
-Its second molten or liquid condition 8, 9, 10, 13, 14, 15, 16, 17, 18, 19, 20, 87, 95, 104, 105, 106, 107, 119, 120, 121, 122, 126, 127, 128.
-Its third solid state 13, 14, 15, 16, 18, 20, 21, 87, 96, 104, 105, 106, 120, 121, 122, 126, 127, 128.
-It cooled down and acquired a condition compatable with the existence of life 109.
Plants-remains of marine plants or fucoids 110.

-the fossile plants of the Coal Era 110.
-of the Devonian Period 110.
-of the Silurian Epoch 110.
-the algæ, &c., that began their life as far back as the Azoic Age, 110.

Play-fair 95.

Pleiades 27, 135 (*vide* Kri-ttikâs).

Pleistocene Age 31.
-Epoch 146.
-Period 63, 138, 141, 143.

Pliocene Period 32.
-Lower-32.
-Man 32.

Pococke 152 (India in Greece).

Polar religions 153.
-Characteristics of the-72.

Post-Tertiary Age 31.

Post-Vedic Geological Literature-cause of the absence **of-7.**

Post-Vindhyan (that is, Cambrian) Epoch 85.

Praja'pati-the Creator 14, 18, 116.

Pralaya or Deluge 141.

Pre-Cambrian or Vindhyan Age 102, 111.
-Epoch 114, 153.
-Period 11, 113.
fossils 129.
-life 111.

Precession of the equinoxes 134.

Precursor of Mankind-the Man-Lion, the connecting link between the brute intelligence and the reasoning powers of man 88, 89.

Pre-Glacial man 32.

Primary Era 21-rocks 93
-use of the term Primary Era by Sir Charles Lyell 21.

Primates 89, 90.

Primitive Ancestors *p.* IX, 50, 52, 53, 56, 57, 58, 59, 60, 106, 138, 148.
-the oldest ancestors of our Rig-Vedic Bards 49.
-of Rig-Vedic Fore-fathers 52, 138.
-A'ryan Ancestors 60.
-A'ryans 59.
-A'ryan Progenitors 61.
-A'ryan Home 144.
-Cradle 144.
-Fore-fathers 49, 72, 76.
-observers of Rain, Dawn, and Sun 46, 49, 53.
-Vedic life 29.

Progenitor Manu 59.

Prometheus 140.

Prophetic announcement 81.

Proterozoic Epoch 115.
-fauna 113.
-flora 113.
-Period 108, 112.

Punja'b 56, 152.

Pura'nas *p*. V.
-Antiquity of the-10, 11.
-Geological references in the-*p*. V, 9, 11,
Pura'na Era 154.
Pura'nic Geologists 88.
Pura'nic Geological works or treatises-cause of the absence of-*pp*. V, VIII, 7.
Pura'nic literature 86.
Pyrrha 140.

Q.

Quaternary Age 148, 156.
-Epoch *p*. IV.
-Era *p*. IX, 26, 33, 132, 133, 134, 143.
-Period 30, 33, 83, 117, 157.
-the period of the R*i*g-Veda extending beyond the Quaternary Epoch, that is, preceding the Great Ice Age and reaching the Tertiary Era 26.

R.

Ragozin 45.
-His work *The Vedic India* 45.
Rain-the fall of-Caused by killing Vr*i*tra-Serpent 49.
-the first showers of-observed by the Primitive Ancestors of our R*i*g.-Vedic Fore-fathers 50, 56, 57.
-the first fall of-or rather the first observatior of the fall of-50, 52, 57.
-clouds of-50.
-the region where they were seen, and the rain-waters had fallen 51, 52.
-waters in the clouds held off by force by Vr*i*tra-Serpent, and therefore the latter killed by Indra. 48, 49, 50, 51, 52, 58, 59.
Ra'jastha'n. Tod's History *p*. V.
Rangha'-the river-identified with the Vedic river Rasâ -78.
Rasa-a Vedic River corrupted into the Zendic Ranghâ 78.
Raspi 97.
Ra'vi-the Vedic river Parushn*i* or Irâvatî 78.
Recent Period 134, 157. (*vide* Quaterrary Era).
Refrigeration 86, 95.
Region-God created-38.
-the scene of vitality 101.
-of primitive vitality 102.
-where life had first commenced 101, and had originated 102, 103.
-where the most ancient life-types are found 101.
Repositories of knowledge-our ancient fore-fathers *p*. I.
Reptile life-types 115.

Researches--Geological—of the East *p.* IX, 8, 9, 10, 11, 12, 13, 14, 15, 16, 17, 18, 19, 20, 21, 22, 23, 86, 87, 88, 89, 90, 91, 97, 99, 100, 101, 102, 103, 104, 105, 106, 107, 108, 109, 110, 111, 112, 113, 114, 115, 116, 117, 118, 119, 120, 121, 122, 126, 127, 128, 129, 130, 131, 132, 133, 135, 136, 137, 138, 139, 140, 141, 142, 143, 144, 145, 146, 147, 148, 149, 155, 156, 157.
-of the West—*p.* IX, 84, 85, 86, 92, 93, 94, 95, 96, 97, 110, 111, 113, 114, 115, 117, 124, 125, 126, 129, 130, 133, 134, 139, 140, 153, 154, 155.

Rhinoceros-remains of-32.
-of the Lower Pliocene 32.

Rhizopod 129.

Rig-Veda *p.* II, III, 31.
-the great antiquity of the-28, 83.
-The most ancient book of the A'ryan world 64.
-the most ancient document in the world 83.
-the spring-heads of thought, language, and of the Poetry of India 64.
-the Rig-Veda and the Vedic literature belong to a period extending from 2500 B.C. to the Tertiary Epoch *p.* II.
-Geological knowledge of the Rishis of the Rig-Veda period 12, 25.
-Geological references in the-19, 20, 21, 22, 23, 100, 101, 104, 105, 108, 109, 111, 112, 113, 119, 120, 127, 128, 130.
-A'ryan speech preserved in the-150.

Rig-Vedic Ancestors autochthonous in A'ryâvarta 82.

Rig-Vedic Ancestors 72, 83.
-Bards 49, 83.
-Fore-fathers 46, 52, 108.
-Poets 52, 56.
-Rishis *p.* IV, 64, 83, 91, 99, 126.

Rishis of the Rig-Vedic times and Tertiary Period 126.

Rivers (*vide* Seven Rivers-Land of the-)

Rocks 2, 3, 93, 94.
-Alluvial 94.
-Alluvium 93.
-Aqueous 93.
-Hypogene 93.
-Metamorphic 93.
-Plutonic 93.
-Primitive 93.
-Secondary 93, 94.
-Stratified 93.
-Transitional 94.

-Transition-formation (class of rocks) 94.
-Unstratified 94.
Rome 152.
Russia' 147.

S.

Sacred Books of the East *p*. V.
Sacrificers-Indra giving ready assistance to the-42.
Sacrifice loving A'ryans-the Brâhmaṇs 58, 59, 60.
-Mortals 59.
Salt Range of the Punjâb 102, 153, 154.
Sanskrit-a wonderful language *p*. I.
-a spoken language 150, 151.
-literature *p*. III, IV, 141.
Sapta-Sindhus (सप्तसिंधवः) *p*. III, 34, 51, 75, 76, 77, 78, 153-the Land of the Seven Rivers of A'ryâvarta.
Sarasvati-the most sacred River of A'ryâvarta; and the scene where life had first commenced 34, 35, 38, 60, 73, 75, 78, 99, 100.
-the slayer of Vritra-(वृत्रघ्नी) 60.
Savants *p*. X, 84, 122.
Sayana 41, 51, 57, 58, 123, 145.
Scene of Creation 38, 75.
-primitive-39.
-of primitive civilization 151.
-of primitive vitality 1 2.
-where life had first commenced 34, 101.
Schist 96, 120.
Schlegel-his opinion as regards the age of the law-giver Manu 8.
Scilla 97.
Science of Geology in the Occident and the Orient 156, 157.
-scope of the-1, 2, 3, 4.
Sea 96, the boiling sea 96.
-weeds 110.
Season-Characteric 71.
-mention of the-in the Taittirîya Araṇyaka 71.
Second deed of heroism of Indra 52, 53 (*vide* the first heroic deed and the third heroic deed of heroism of Indra).
Secondary or Mesozoic Epoch 108.
-Period 22.
-rocks 93, 94.
Sediments 16, 96.
Sequence of Evolution 25.
Serpent or Ahi (अहि) 49, 50, 51, 52, 58, 59.
-Clouds-like 49.
-the first born of-47, 48, 50.
-the first killed of-50, 52.
-the greatest of-48, 50.

-the oldest of-48, 50.
-slayer of the-(वृत्र Vritra serpent)-*viz.* the Indra 42, 43, 47, 49, 50, 53.
Settlements (or colonies) in the A'rctic regions of us-the Indo A'ryans-35, 72, 73.
Seven Rivers 51, 52, 59, 60, 75, 76, 78, 138.
-the Land of the-36, 39, 40, 46, 48, 56, 57, 58, 61, 73, 75, 76, 77, 139, 149, 153, 154.
Sharyana'vat Lake in the Province of Kurukshetra 42.
Shata-Patha-Brâhmaṇa 27, 77, 88, 113, 135, 136, 139, 140, 141, 142, 150.
Shiwa'lik Mammalia 31.
Shiwa'lik Yuga (शिवालिक-युग) 11.
Shropshire-the oldest known fossil fish found in-129.
Shutudri-a Vedic river-the modern Satlaj 78.
Sia'm 152.
Silurian Epoch 110, 154.
-flora 110.
-fossils 102, 153.
-group of the Upper-129.
Siluro—Cambrian (Ordovician) flora 110.
Simpler conceptions-progress from-to more highly organised orders 96.
Sindhu 52, 78. (*vide* the Indus).
Snow 140, 144, 148, 155.
-floods of-141.
-sheets of-148, 155.
Snowy Peak 138.
Solid state of Earth 13, 14, 15, 16, 18, 20, 21, 96, 104, 105, 106, 120, 121, 122, 126, 127, 128.
Soma 39, 40, 41, 42, 43, 44, 45, 46, 47, 59.
-antiquity of-43.
-drink of-42.
-juice of-47.
-seat of the worship of-45.
-the place of birth and growth of-39, 40, 41, 42.
Southern hemisphere-similarity of Silurian fossils in the-to those in the northern 153.
Spencer J.W.-Professor-85.
Spiegel 44.
Spinoza 91.
Stars rising and setting of the-seen only once a year in the Arctic regions 80.
Startling discoveries in Geology, made by our Rig-Vedic and Vedic Forefathers 91.
Steno 92, 97.
Suma'tra' 152.
Sun 39, 40, 49, 52, 53, 54, 55, 56, 64, 65, 68.

Surya Siddhânta–an astronomical work of Bhâskarâchârya 71, 72.
Sutledge–the river Shutud-i of the Vedic and Purâṇic times 56, 78 (*vide* Shutudri).
Swa'mi Vidyâraṇya 123.

T.

Taittiriya A'raṇyaka–Geological references in–14, 18, 70, 71, 87, 88.
Taittiriya Brâhmaṇa–Geological references in–13, 14, 15, 17, 18, 19, 20, 27, 28, 70, 87, 105, 106, 116, 119, 120, 129.
Taittiriya Samhitâ–Geological references in–13, 67, 68.
Taittiriya Upanishad–Geological references in–14, 15, 17, 18, 87, 88, 105, 111, 112, 115, 119, 120 127, 128.
Taylor–Dr. Isaac 45, 151.
–His Work *The Origin of the A'ryas* 45, 151.
Tertiary Ancestors 67, 69.
–Fore–fathers 68.
–Framers of the most wonderful language–the Sanskrit *p*. I. (*vide* the Oldest Ancestors).
Tertiary Epoch *p*. II, 26, 32, 33, 34, 63, 72, 82, 83, 84, 103, 108, 109, 112, 113, 133 (*vide* Cainozoic Epoch).
–Era 12, 29, 83, 108.
–Period *p*. I, 5, 12, 34, 80, 126.
–Hymns and utterances of the Rig-Vedic Rishis, during the–26, 83.
–Rig-Vedic period part of the Tertiary Era 12.
–Vedic Rishis of the–5, 12, 33, 121.
–Vedic Ancestors of the–34.
Tertiary Mammalia 5, 10, 14, 21, 108.
Tertiary Man 5, 6, 12, 32, 33, 34, 83, 108, 109, 112, 113.
Thermal Ocean 16, 21, 87, 95, (*deep*) 105, 106, 119, 120.
Thessaly 140.
Third deed of heroism of Indra 52, 53. (*vide* the First heroic deed and the Second deed of heroism of Indra.) 52, 53.
Thornton 98.
Thunderbolt 40, 47, 48. (*Vide* Vajra वज्र)
Tilak–Mr. Bâl Gangâdhar 26, 28, 30, 141.
–His interesting Works:—
(1) The Arctic Home in the Vedas, and

(2) The Orion. 27, 28.
-His views in regard to the antiquity of the Vedas 19, 26, 27, 28, 135, 141.
Tishya Star 27, 28.
Tod-Colonel-His historical work *The Annals of Rája-sthán* *p.* V.
Tortoise 18, 88, 115.
-Mesozoic-14.
Transition-use of the term-94.
-formation 94.
Treta (Yuga or Age) 143, 144, 145, 146, 156.
Trias 115.
Trilobites-life in 87, 108, 111, 128.
Tropics 153.

U.

Upanishads-Geological references in the-12, 13, 14, 15, 23, 89, 124, 127.
-Taittirîya-15, 88, 105, 111, 112, 115, 127, 128, 130.
-Nrisimha Purva T. Upanishad 89, 116.
-Chhandogyopanishad 124.
-Approximate date of the-26.
Ushas-(the Dawn)-Creation of the-by Indra 52, 53, 55.
Ushas disappearing in the Sun, in the region of the river Vipâsh of A'ryâvarta 55, 56.
Ustanvaiti-Zend Gâthâ 79.

V.

Vaivasvata 142.
Vajra (वज्र) 40, 47, 48 (*vide* Thunderbolt).
Vandalism *p.* V, VIII, 7.
Vara (enclosure) 142.
Va'yu Purâna-Geological references in-10.
Vedas *p.* V, VI.
-narration of the Deluge story in the-141, and Vedic works 135.
-unassailable evidence in respect of the advent of the Great Ice Age in the-156.
Vedic acquaintance with Geology *p.* III, VIII.
-Knowledge of Geology *p.* III, 4, 5, 6, 23, 98, 102, 103, 121, 132.
-ancestors 68, 117, 121.
-A'ryans 6.
-evidence (as also Avestic, foreign, and geological), in support of the A'ryan Cradle in A'ryâvarta 155.
-exegetist 123.
-fathers *p.* VIII, IX, 4, 132, 135
-fathers of geology 126.
-fore-fathers *p.* III, 91, 119, 132, 157.
-geologists 88, 115, 156.
-geology 156, 157.

-literature *p*. II, IV, 4, 5, 8, 86, 132.
-Geological references in the Vedic literature 4, 8, 13, 14, 15, 18–25, 86-91, 99–121, 127–157.
Vedic researches in geology 108.
Vedic Rishis *p*. III, 5, 23, 25, 26, 91, 121, 156.
-of the Tertiary period 5, 121, 126.
Vedic times and the cultivation of the science of geology 86.
Vedic Yugas 143.
-their explanation 144,145, 146, 147, 148, 149, 156, 157.
Vendidad 78, 79, 80, 82.
Verethrajao 44 (*vide* Vr*i*trahâ).
Vernal equinox 27.
Vidya'ranya Swâmi 123.
Vincent Smith *p*. V.
Vindhyan Age 111.
-Epoch 114.
-Era 102, 153, 154.
-Period 11, or Pre–Cambrian.
Vipat, Vipash, or Vipâshâ, a Vedic river, and the modern Beeas 40, 55, 56, 57.
Vishnu Purâṇa–Geological references in the–10, 18, 88, 116.
Vital gradations 25, 96.
Vitality on Earth 9, 10, 21, 94, 100, 101, 109, 111.
Vitality in moss 87.
,, in the weeds 87, 88,130.
,, in the lotus 87, 88.
,, in the herbs 87, 88,109. 115, 128, 130.
,, in plant 87, 109.
,, in trilobites 87, 108, 109, 111.
,, in fishes 87, 88, 111, 113, 115, 130.
,, in reptiles 87, 88, 111.
,, in tortoise 88, 112, 115, 130.
,, in boar 88, 130.
,, in Man–Lion or Ape 88, 89, 112, 130.
,, in Man 88, 109, 130.
Vivanhan, a corrupt form of Vaivasvata, appearing in the R*i*g-Veda 142.
Vivasvata 142.
Vritra (वृत्र) 39, 40, 46, 47, 48, 49, 50, 51, 52.
Vritraghni 60 (*vide* Sarasvatî सरस्वती वृत्रघ्नी).
Vritraha' 44.
Vritra Serpent 51 58, 59, 60 (*vide* Serpent).
Vritra Slayer (or Slayer of Vr*i*tra) 49.
Vritra slaying Indra 42, 52, 53, 59.

W.

Waagen 114.

Walcott 85.
Wa'ma Deva R*i*shi–a R*i*g-Vedic poet that had described the Dawn as crushed by the Sun 54–55.
Wa'mana (Incarnation) 88, 112, 116, 131.
Weber–His views in respect of the antiquity of the Vedas 19.
Weeds 87, 88, 113, 128, 130.
Werner 92, 94, 95.
Western Geologists 1, 2, 3, 24, 84, 85, 92, 93, 94, 95, 96, 97, 101, 102, 110, 114, 124, 129, 156.
William Smith 92, 96.
–the father of English geology 96.
Wilson–H. H. Professor 11, 100, 104.

Y.

Yama 142.
Yamuna' 78.
Ya'ska 55.
Year–God 70, 71.
Yima 82, 142. (*vide* Vaivasvata and Vivanhan).
Yuga 145, 156.

Z.

Zarathustra 79.
Zend 79.
–where spoken 150.
Zendavesta 44.
Zeus 139, 140.
Zeus Phyxios 140.
Zoroastrians 40.